DATE DUE

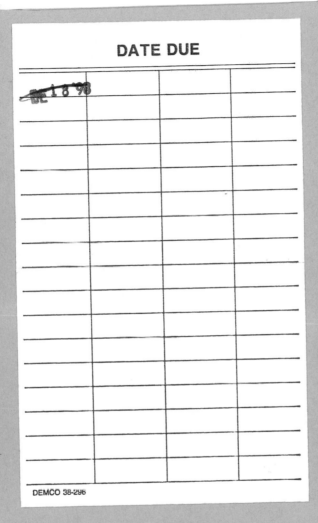

DE 18 98			

DEMCO 38-296

Following the Trail of Light

A Scientific Odyssey

Following the Trail of Light
A Scientific Odyssey

Melvin Calvin

PROFILES, PATHWAYS, AND DREAMS
Autobiographies of Eminent Chemists

Jeffrey I. Seeman, Series Editor

American Chemical Society, Washington, DC 1992

Library of Congress Cataloging-in-Publication Data

Calvin, Melvin, 1911–
  Following the Trail of Light: A Scientific Odyssey/Melvin Calvin
    p. cm.—(Profiles, pathways, and dreams: autobiographies of
eminent chemists, ISSN 1047–8329)
  Includes bibliographical references and index.
  ISBN 0–8412–1828–5 (cloth).—ISBN 0–8412–1829–3 (paper)
  1. Calvin, Melvin, 1911–   2. Chemists—United
States—Biography.   I. Title.   II. Series: Profiles,
pathways, and dreams.
  QD22.C32A3   1992
  540'.92—dc20
  [B]   92–3989
                    CIP
Jeffrey I. Seeman, Series Editor

The paper used in this publication meets the minimum requirements of American
National Standard for Information Sciences—Permanence of Paper for Printed Library
Materials, ANSI Z39.48–1984.

∞

1992 Advisory Board

Foreword

In 1986, the ACS Books Department accepted for publication a collection of autobiographies of organic chemists, to be published in a single volume. However, the authors were much more prolific than the project's editor, Jeffrey I. Seeman, had anticipated, and under his guidance and encouragement, the project took on a life of its own. The original volume evolved into 22 volumes, and the first volume of Profiles, Pathways, and Dreams: Autobiographies of Eminent Chemists was published in 1990. Unlike the original volume, the series was structured to include chemical scientists in all specialties, not just organic chemistry. Our hope is that those who know the authors will be confirmed in their admiration for them, and that those who do not know them will find these eminent scientists a source of inspiration and encouragement, not only in any scientific endeavors, but also in life.

M. Joan Comstock
Head, Books Department
American Chemical Society

Contributors

We thank the following corporations and Herchel Smith for their generous financial support of the series Profiles, Pathways, and Dreams.

Akzo nv

Bachem Inc.

E. I. du Pont de Nemours
and Company

Duphar B.V.

Eisai Co., Ltd.

Fujisawa Pharmaceutical Co., Ltd.

Hoechst Celanese Corporation

Imperial Chemical Industries PLC

Kao Corporation

Mitsui Petrochemical Industries,
Ltd.

The NutraSweet Company

Organon International B.V.

Pergamon Press PLC

Pfizer Inc.

Philip Morris

Quest International

Sandoz Pharmaceuticals
Corporation

Sankyo Company, Ltd.

Schering–Plough Corporation

Shionogi Research Laboratories,
Shionogi & Co., Ltd.

Herchel Smith

Suntory Institute for Bioorganic
Research

Takasago International
Corporation

Takeda Chemical Industries, Ltd.

Unilever Research U.S., Inc.

Profiles, Pathways, and Dreams

Titles in This Series

About the Editor

JEFFREY I. SEEMAN received his B.S. with high honors in 1967 from the Stevens Institute of Technology in Hoboken, New Jersey, and his Ph.D. in organic chemistry in 1971 from the University of California, Berkeley. Following a two-year staff fellowship at the Laboratory of Chemical Physics of the National Institutes of Health in Bethesda, Maryland, he joined the Philip Morris Research Center in Richmond, Virginia, where he is currently a section leader. In 1983–1984, he enjoyed a sabbatical year at the Dyson Perrins Laboratory in Oxford, England,
and claims to have visited more than 90% of the castles in England, Wales, and Scotland.

Seeman's 80 published papers include research in the areas of photochemistry, nicotine and tobacco alkaloid chemistry and synthesis, conformational analysis, pyrolysis chemistry, organotransition metal chemistry, the use of cyclodextrins for chiral recognition, and structure–activity relationships in olfaction. He was a plenary lecturer at the Eighth IUPAC Conference on Physical Organic Chemistry held in Tokyo in 1986 and has been an invited lecturer at numerous scientific meetings and universities. Currently, Seeman serves on the Petroleum Research Fund Advisory Board. He continues to count Nero Wolfe and Archie Goodwin among his best friends.

Contents

Photographs

Preface

"HOW DID YOU GET THE IDEA—and the good fortune—to convince 22 world-famous chemists to write their autobiographies?" This question has been asked of me, in these or similar words, frequently over the past several years. I hope to explain in this preface how the project came about, how the contributors were chosen, what the editorial ground rules were, what was the editorial context in which these scientists wrote their stories, and the answers to related issues. Furthermore, several authors specifically requested that the project's boundary conditions be known.

As I was preparing an article[1] for *Chemical Reviews* on the Curtin–Hammett principle, I became interested in the people who did the work and the human side of the scientific developments. I am a chemist, and I also have a deep appreciation of history, especially in the sense of individual accomplishments. Readers' responses to the historical section of that review encouraged me to take an active interest in the history of chemistry. The concept for Profiles, Pathways, and Dreams resulted from that interest.

My goal for Profiles was to document the development of modern organic chemistry by having individual chemists discuss their roles in this development. Authors were not chosen to represent my choice of the world's "best" organic chemists, as one might choose the "baseball all-star team of the century". Such an attempt would be foolish: Even the selection committees for the Nobel prizes do not make their decisions on such a premise.

The selection criteria were numerous. Each individual had to have made seminal contributions to organic chemistry over a multidecade career. (The average age of the authors is over 70!) Profiles would represent scientists born and professionally productive in different countries. (Chemistry in 13 countries is detailed.) Taken together, these individuals were to have conducted research in nearly all sub-specialties of organic chemistry. Invitations to contribute were based on solicited advice and on recommendations of chemists from five continents, including nearly all of the contributors. The final assemblage was selected entirely and exclusively by me. Not all who were invited chose to participate, and not all who should have been invited could be asked.

A very detailed four-page document was sent to the contributors, in which they were informed that the objectives of the series were

1. to delineate the overall scientific development of organic chemistry during the past 30–40 years, a period during which this field has dramatically changed and matured;

2. to describe the development of specific areas of organic chemistry; to highlight the crucial discoveries and to examine the impact they have had on the continuing development in the field;

3. to focus attention on the research of some of the seminal contributors to organic chemistry; to indicate how their research programs progressed over a 20–40-year period; and

4. to provide a documented source for individuals interested in the hows and whys of the development of modern organic chemistry.

One noted scientist explained his refusal to contribute a volume by saying, in part, that "it is extraordinarily difficult to write in good taste about oneself. Only if one can manage a humorous and light touch does it come off well. Naturally, I would like to place my work in what I consider its true scientific perspective, but . . ."

Each autobiography reflects the author's science, his lifestyle, and the style of his research. Naturally, the volumes are not uniform, although each author attempted to follow the guidelines. "To write in good taste" was not an objective of the series. On the contrary, the authors were specifically requested not to write a review article of their field, but to detail their own research accomplishments. To the extent that this instruction was followed and the result is not "in good taste", then these are criticisms that I, as editor, must bear, not the writer.

xviii

As in any project, I have a few regrets. It is truly sad that Egbert Havinga and Herman Mark, who each wrote a volume, and David Ginsburg, who translated another, died during the course of this project. There have been many rewards, some of which are documented in my personal account of this project, entitled "Extracting the Essence: Adventures of an Editor" published in *CHEMTECH*.[2]

Acknowledgments

I join the entire scientific community in offering each author unbounded thanks. I thank their families and their secretaries for their contributions. Furthermore, I thank numerous chemists for reading and reviewing the autobiographies, for lending photographs, for sharing information, and for providing each of the authors and me the encouragement to proceed in a project that was far more costly in time and energy than any of us had anticipated.

I thank my employer, Philip Morris USA, and J. Charles, R. N. Ferguson, K. Houghton, and W. F. Kuhn, for without their support Profiles, Pathways, and Dreams could not have been. I thank ACS Books, and in particular, Robin Giroux (production manager, formerly acquisitions editor), Janet Dodd (senior editor), Joan Comstock (department head), and their staff for their hard work, dedication, and support. Each reader no doubt joins me in thanking 24 corporations and Herchel Smith for financial support for the project.

I thank my children, Jonathan and Brooke, for their patience and understanding; remarkably, I have been working on Profiles for more than half of their lives—probably the only half that they can remember! Finally, I again thank all those mentioned and especially my family, friends, colleagues, and the 22 authors for allowing me to share this experience with them.

JEFFREY I. SEEMAN
Philip Morris Research Center
Richmond, VA 23234

April 7, 1992

[1] Seeman, J. I. *Chem. Rev.* **1983**, *83*, 83–134.
[2] Seeman, J. I. *CHEMTECH* **1990**, 20(2), 86–90.

Editor's Note

AS A YOUNG GRADUATE STUDENT IN THE LATE 1960s, I was impressed with the four Halls—Gilman, Hildebrand, Latimer, and Lewis—that formed a quadrangle within the chemistry complex at Berkeley. I knew well the name of G. N. Lewis; the fact that Joel Hildebrand, while still alive and doing research, had had a building named after him, was most impressive of all. The Berkeley pride in its association with these seminal scientists was very evident!

In 1964, a new building was added to the complex, but it was separate from the others in more ways than one. Located some 500 yards away, across a stream and past the faculty club, the structure was also set apart because of its unique design it was round! Dubbed the "Round House"*, it was also referred to by some as "Calvin's Carousel", although officially it was the Laboratory of Chemical Biodynamics.

Those of us in the quadrangle were aware that the Round House was part of the chemistry department and was Nobel laureate Melvin Calvin's domain, but few of us had had the opportunity to work within that unique structure, nor to become commingled with those who did. True, a few grad students from the Round House showed up for the required courses, but immediately thereafter they, and we, disappeared into our respective labs. Calvin was seldom seen in our arena, and most of us never visited his, but such was pure chemistry— the highest purpose in life—or so we thought then. How parochial we were!

The Round House reflected Calvin's style, both professionally and personally. Inside, the laboratory was very open and airy, with many windows and few walls. Like its director, it encouraged interaction among the inhabitants. In fact, there is a now-legendary coffee table that was a focal point for famous guests and exciting ideas. Not surprisingly, the men and women on his staff were very bright and very enthusiastic. People were drawn to his lab by the promise that they could expand their scientific horizons in a milieu of multiple disciplines.

It is likely that two distinct features mutually supported the unique multidisciplinary nature of Calvin's laboratory. Calvin's scientific inclinations were certainly broad and far-encompassing. In addition, the physical separation, first of the Old Radiation Laboratory, and later, the Round House, from the other chemistry buildings added compelling strength and cohesiveness to the members of his Bio-Organic Group.

"I was always part of the departmental activities," Calvin recalls. "I participated in meetings and regularly attended the lunches at the chemistry table at the faculty club. I maintained an office in Latimer Hall. But having our group in one facility, distant from the other buildings, simply minimized distractions. Being separate wasn't the result of a conscious effort, but rather just the way it happened. With the demolition of the Old Radiation lab, had the Round House not been constructed, we would have consolidated elsewhere. We certainly had to face less bureaucracy being in another building." Perhaps the old adage "out of sight, out of mind" best describes the situation.

Why was it, and why is it, that university research seems dominated by the single-discipline approach? I have searched my memories for even subtle restrictions on us as grad students that would have strongly influenced me in that direction. Were we too busy, too focused? Why was there so little interaction among organic chemists and physical chemists? I do recall undercurrents of competition for resources and for space. Could tenure (or reputation) be achieved by collaborations outside one's field, with other senior investigators as coauthors?

Ironically, a strong case can be made that many of the greatest scientific discoveries of the past few decades have been—and in the future will be—multidisciplinary in nature. Melvin Calvin clearly has been a leader in pursuing an intellectual concept to wherever it might go and in bringing together whatever resources have been required to follow the "Trail of Light". He assembled scientists whose disciplines ranged from psychology to botany, from organic photochemistry to chemical evolution. According to a former student, "Traditional boundaries of scientific disciplines disappeared. The breadth and scope of science in the Calvin laboratories were incredible. People from all over the world worked side by side . . ."

In Calvin's words:

There is no such thing as pure science. By this I mean that physics impinges on astronomy, on the one hand, and chemistry and biology, on the other. The synthesis of a really new conception requires some sort of "union in one mind" of the pertinent aspects of several disciplines.

Indeed, Melvin Calvin has revealed a strong philosophical bent to his nature. Over the years, his friends and colleagues have collected numerous Calvin maxims: "One should do what one wants to do"; "Don't be afraid of ideas, either your own or others"; "It is better to act in relative ignorance than not to act at all"; "We are here to learn, and as chemists we can learn better than most"; "Follow the research wherever it takes you"; "Have confidence not to fear the other scientific disciplines and their mysteries". These are the words of a leader and an explorer. More are scattered throughout this volume.

In fact, my own daughter was the recipient of a bit of Calvin wisdom. Last year, my then-12-year-old daughter, Brooke, called me at the lab, all excited. Her science teacher had given the class an assignment to write a paper on one of 10 scientists, and Melvin Calvin's name was on the list. "Isn't he one of your authors?" Brooke asked. When I confirmed that he was, she indicated that she would like to write to him. "I want his autograph," she explained. "No," I said, "you have to ask him a question, and then, with luck, you'll get both a reply *and* his autograph!" So after much thought, she decided to ask Calvin which was more memorable to him, his science or his Nobel Prize. Within a week Calvin responded:

> I suppose the simplest way to answer is to tell you that there is nothing, in my life at least, that surpasses the pleasure which a successful scientific activity gives. Everything else is peripheral to that.

Following the Trail of Light . . . the ancient peoples, and many since, have worshiped the sun and its powers. Light is that eternal substance of life, clarity, richness, and fullness. Melvin Calvin has been one of science's greatest examples, stepping out from its shadow, basking in its strength, and following its reflection to wherever it led.

JEFFREY I. SEEMAN
Philip Morris Research Center
Richmond, VA 23234

May 26, 1992

---

*In 1980, upon Calvin's retirement as the Director of the Laboratory of Chemical Dynamics and elevation to University Professor Emeritus, the Round House was officially renamed the Melvin Calvin Laboratory. As had Hildebrand, so too has Calvin experienced the unique pleasure of being so honored during his lifetime.

Following the Trail of Light

A Scientific Odyssey

Melvin Calvin

The Childhood Years

My father, Elias Calvin, came from Kalvaria in what is now Lithuania, arriving in New York and making his way to the Midwest. In fact, according to my father, the name Calvin originated with the immigration authorities and was based on his birthplace in Europe. My mother, Rose (Hervitz) Calvin, came to the United States from the Georgian region of Russia.

By the time I was born in 1911, my parents were living in St. Paul, Minnesota, in a middle-class neighborhood in the typical milieu of that era of the 20th century. My sister, Sandra, was born several years later, and the two of us lived and played in the St. Paul–Minneapolis area with our cousins (my father's brother also lived in St. Paul) and were involved in the usual family activities and outings. (Incidentally, some of the sons and daughters of these cousins are now living in the Bay Area of San Francisco, and we see each other from time to time to maintain the family connection.)

My father was a cigar maker in St. Paul and only changed professions when there was a spurt of activity in the automobile industry. At that time we moved to Detroit, where he was employed by the Cadillac Motor Car Company. Neither of my parents went to college. However, my father was a very talented auto mechanic and this gave him great pleasure; I believe he was very happy in his work at that time. My mother was always talented with her needle, and throughout her life

In Indian garb.   *In the back yard in Minnesota.*

made a very good living in designing and creating distinctive articles of clothing, household decorative fabrics, and specially knitted materials for both the wholesale and retail trade. (Later, when my mother was living in Southern California near my sister, she had her own flourishing business which she continued until she was in her seventies, and quite successfully.)

The photographs reproduced here are some of the oldest in the family archives. I don't quite remember how old I must have been when I was an Indian, but I do remember the horse in the back yard. That was one of the things that we did in that era in that part of the country. It is strange to realize that now places like my neighborhood in St. Paul are part of a large city, and not the quiet, countrylike streets that allowed us as children to roam about the city and its environs. The picture of me and my sister, Sandra, is one of the few I have of the two of us together.

Apparently, even at age 3 I showed an intense desire to learn. While watching my uncle, a university student, studying, I decided to "study" too, so I went to my own room, where my grandmother found me pacing up and down. When she asked me what I was doing, I answered her that I was studying.

According to my family, I used to dismantle all of my toys, and as soon as I understood how and why they worked, I would put them together again.

When we moved from St. Paul to Detroit, I was old enough to begin high school, and I attended Central High School (now the site of Wayne State University). In high school I really didn't have a very clear distinction in my mind between science and technology. Because I was curious about the nature of things, I called it all "science". One of my most exciting moments over the years occurred during my first year as a high school student as I learned the nature of the periodic table and its construction according to the hydrogenic principles of the Bohr atom. The realization that all of the elements were made up of the same units of positive charge and electrons and they were chemically different only by the difference of the number of electrons they had circulating about the heavy positive nuclei was to me a revelation of personal discovery in the mid-1920s (I think some time in 1926). If this particular idea and its significance were expressed in the high school textbooks of the day, I don't remember. It seemed to me a personal discovery!

With sister, Sandra.

My sister, Sandra, was also growing up in a typical midwest family pattern. Unfortunately, because of the economic climate of the times, she was not able to go to college. The resources were only enough for me to continue my education. I have always felt sorry about this, as Sandra is a very bright woman and undoubtedly would have benefited in her life from the opportunity for higher education.

Education

High School Years, 1924–1927

I should like to step back in time a bit to the last several years of my high school days in Detroit, when my best friend, Abraham Becker, who was first in our high school class, and several other of our friends, and I were members of a birding club sponsored by the Detroit Department of Recreation. This birding club was part of a garden group in central Detroit, which had its gardens in a vacant lot just across from what is now the Detroit Public Library. The garden teacher, Miss Ives, undertook to form a bird club with some of us who had gardens there and to take us on bird walks wherever she could. These places were usually around the edges of town, particularly in the cemeteries, which were the only green places we could reach very easily.

During this period we became thoroughly familiar with the birds of the eastern United States, at least those that migrated through lower Michigan. Our task was to recognize the birds on sight or by sound if possible. This early experience has remained with me to this day. I keep track of all the birds I see, particularly now in California, and anywhere else that we travel. Of course, the birds of the West Coast, although analogous to those of the eastern United States, are different in many respects. For example, the jays we have here in the West are not the blue jay of the East but the Canada jay and the Steller's jay. They are easy to recognize as jays but, nevertheless, different from the eastern ones. The same thing holds true for many other birds with which we became familiar in the East, and I had to learn about birds all over again here in California. It wasn't a great task because they are recognizable, even though the individual species or variety might be slightly different in the West. This early introduction to the birds has provided a substantial source of pleasure on our travels, and especially on our ranch in northern California.

Abe Becker and I performed our first scientific experiment in biology at about 12 years of age. I lived in a flat over the family grocery store; his family had a house a block or so from where I lived, with a nice back yard in which we frequently played. One afternoon we were chasing grasshoppers in the yard. We caught one. There was a little pan of water nearby. We thought we would try to see if the grasshopper could swim, so we put him in the pan. He did swim around, kicking, and we then wondered what would happen if we pushed him under the water. We did this and his kicking ceased. We fished him out and laid him on a piece of paper in the sun to dry. In a few minutes he awoke and hopped away. This was mysterious to us. Here was an insect that had apparently drowned and was fully able to recover, merely by drying out! I still wonder about this phenomenon and can only rationalize it today in terms of oxygen deprivation leading to temporary deprivation of nerve and muscle function. But the grasshopper, unlike the human, does not lose nerve function merely upon oxygen deprivation for a short time.

As a high school student I had taken a job in what was then a "supermarket". At that time these grocery stores were not what the establishments are today, but for that day and age, the store I worked in was a large one. During the few moments that I had to myself in such a very active job, I realized how dependent was everything I did in such a place on the function of chemists and chemistry. The composition of foods (which was a very sophisticated subject to me then), the paper around the cans, the dyestuffs in the paper for printing the labels and the boxes, the cans themselves, the paper boxes—all of this convinced me that chemistry was an important component of our daily lives. The thing that I was concerned about was the need to be assured of an adult gainful occupation.

My concern stemmed from the fact that my father had had some difficulties. Upon his arrival in this country from Russia as an immigrant, he became a cigar maker, but that employment came to an end, and he was required to find other ways of supporting his family. About that time there was a burst of employment in the automobile factories in Detroit, so the family moved from St. Paul, Minnesota, to Detroit. My father obtained employment there and became a skilled mechanic in the Cadillac automobile factory. He rose as far as one could in such a position because of his skill, and he was very proud of it.

Eventually my family saved up enough money to start a "mom and pop" grocery store, and I began learning about that business while I was still in grade school. Thus, when I entered high school and sought weekend work, I went to a large market. This, in turn, gave rise to my decision to enter chemistry as a profession.

High school graduation photo.

Undergraduate Education, 1927–1931

My undergraduate career began as a freshman at the Michigan College of Mining and Technology (now Michigan Technological University) in 1927. My attendance at this college was somewhat of an accident. At that time the institution was just emerging from its period as a mining school and wanted to expand its student population to include all kinds of technology and science. In order to do this it had established a scholarship that was available to every high school in the state of Michigan. The high school was to select its highest-standing graduate, who would be eligible for such a scholarship. It so happened that I was the second-standing student in my graduating class at Central High School in Detroit. Fortunately my best friend, Abraham Becker, who was the first-standing student in that class, had already committed himself to medical studies at the University of Michigan. (He has been a practicing physician (internist) in Detroit and in Birmingham, Michigan, ever since. I have seen him many, many times in the intervening years and periodically talk to him on the telephone.) Abe was not available because of his plans to attend medical school, so it fell to me to accept the scholarship. It was appropriate because I had already decided on a career in chemistry.

In the autumn of 1927 I took the night train from Detroit and arrived in Houghton (on the northern peninsula of Michigan) a day or two later. It turned out to be a very pleasant and, I think, fortunate 4 years for me. Because I was their first student in chemistry, their chemical curriculum was rather limited. I learned my freshman, analytical, and inorganic chemistry from Professor Bart Park, and I learned my organic chemistry and advanced physical chemistry from Professor Charles M. Carson. These were very straightforward courses for their time, and I had no idea of where the frontiers of chemistry were at that time. In fact, because there was a very limited curriculum in chemistry, I found that it was necessary for me to fill in my curriculum with other courses, primarily mineralogy, geology, and paleontology. I even took a summer course in civil engineering, which I found very interesting. The effect of this nonchemical undergraduate period has never left me, and I am very happy (in fact, I think I was very fortunate) to have such a diverse experience.

It is very difficult for me to remember precisely what was happening in science at the time of my undergraduate years, 1927–1931. The organic chemistry must have been very straightforward structural and some elementary synthetic concepts. I do remember the classical physical chemistry studies, such as measuring a molecular weight by depression of a freezing point and other such standard measurements. At that time, of course, we had no such equipment as automatic spectrometers to take absorption spectra for us. What few spectra we did see were either line spectra taken on photographic plates or general continua, which were hardly useful.

I found it necessary for financial reasons to take a year off from my undergraduate education to earn some money after my sophomore year. I returned to Detroit, where my family was, and found a job as an analyst in a brass factory (Revere Copper and Brass Company at that time). I worked the swing shift, from 4 p.m. to midnight. During the day I took some classes at what was then City College of Detroit (now Wayne State University). I carefully avoided taking any scientific courses, and I took a course in world history and another in philosophy. After going to school I went home and then drove a very old Oldsmobile to work, which was far down on Jefferson Avenue near the waterfront on the Detroit River.

My work began with a walk through the foundry, in which the brass was being manufactured. I removed a ladle of the molten brass as a sample. This turned out to be a fraction of a sphere about 3 inches in diameter. I collected and marked the samples and took them back to the laboratory. My other collection involved taking samples of finished rods and sheets. All the samples were carefully marked and brought back to the laboratory.

I then did my analyses, which consisted of dissolving brass shavings from each of the samples in nitric acid and then adding ammonia to the beaker until the solution became basic. Each beaker was about 250 mL and contained about 25–50 mL of solution. At this point the beakers were lined up in a hood. After allowing them to cool, I would shake them gently, at which point the traces of iron present in the brass would coagulate in the form of iron hydroxide. I learned to estimate how much trace iron was in the sample by the amount of hydroxide I could see. Having done that, I filtered off the ferric hydroxide and did a thiosulfate titration for copper. At the end of the shift I turned in my sheets with these numbers on them. During the day shift the head of the laboratory would then decide the fate of the batches from the analytical results. This job occurred after my freshman year when I had completed freshman chemistry, so I knew a little bit about what I was doing. After completing this year away from Houghton, I returned and finished my B.S. degree work.

The only course in biology I have ever taken is represented by the paleontology I took during my undergraduate years at Michigan Tech. It was, of course, largely devoted to those materials that would fossilize (that is, bones and hard parts of organisms), and it provided my introduction to evolution as well as to some of the Darwinian theories of how it took place. This background led much later to our work on chemical evolution and the origin of life, as well as some aspects of our organic geochemical studies, which I have called molecular paleontology in contrast to the paleontology that I studied as an undergraduate.

The fact that this was all the biology I ever formally studied appears to me to have been an advantage. I was quite definitely trained in the fundamental unifying concept of biology, namely, evolution and its various theoretical mechanisms, but I was not encumbered with preconceived notions of how the actual biochemistry of living systems worked. I learned that perforce as I went along in my research in later years. I have even learned a good bit of taxonomic classification, but I have difficulty even today analyzing the fundamentals of structure upon which today's subdivisions are based. I have found myself asking professional biologists embarrassing questions about the basis of such taxonomic differences that they take for granted and that do not appear obvious to me. They, in turn, find themselves reexamining the basis of such differences. Sometimes they can answer my questions and sometimes they cannot.

The other aspect of this early training that still makes its appearance very frequently in our current activities is the result of the course in mineralogy taught by the son of the professor of geology and paleontology. This course was taught in a very useful way, at least for me. On each occasion, young Professor Seaman would bring to the class a tray of samples of minerals for each student. The task was to identify those minerals as closely as possible within the time period of the laboratory afternoon. It was kind of a game, in a way, which I enjoyed very much. The result is that when my wife, Genevieve, and I traveled in the mountains or in any other region in which rocks are exposed, her tendency was to pick up those that are especially beautiful, and my look at them was always in terms of the mineralogy they might provide. This was not a source of any conflict, but rather a mutually appreciated bit of knowledge.

Graduate Education, 1931–1935

In the beginning of my fourth year at Houghton it became clear that I would seek to undertake graduate work in chemistry and that I should begin my applications to various graduate departments before the end of the 1930 academic year. This I did. I

S. C. Lind, chairman of the Department of Chemistry, University of Minnesota.

don't remember all the schools to which I applied, but I received positive responses from several of them, including the University of Minnesota in Minneapolis. I accepted this one, primarily because by that time my mother and father had moved back from Detroit to the Twin Cities, and it was obvious that was where I should go to school as a graduate student.

In the fall of 1931, when I entered the University of Minnesota as a graduate student, Professor Samuel C. Lind was the chairman of the Department of Chemistry. He had recently arrived there from the U.S. Bureau of Mines and had brought with him the United States' supply of radium, which he kept in a lead-lined safe somewhere in the basement of the chemistry building. I learned later that the way he used the material was to withdraw radon gas from it through a glass conveying tube to the outside of the safe. He could fill small thin glass-walled spherical containers, about 1–2 mm in diameter, on the end of a capillary tube, with the radon, which was given off from the solid salts of radium that he kept in the safe. This radon provided a source of alpha particles and could be used to study the effects of alpha particles on materials, particularly on chemical reactions.

During that first few months in Minneapolis I was very much tempted to undertake research in this area, but for some reason I did not commit myself. Instead, I undertook a graduate association with Professor George Glockler, who offered me a position as graduate advisor without any obligation to do work in his area. That temporary association became a permanent one by the end of that first year, when I began to work on the electron affinity of halogen atoms, particularly of iodine.[1] This work took me almost 4 years and was expanded into two of the other halogens, bromine and chlorine.

About the time I was ready to write my thesis, another chemist, Joe Mayer, published some measurements on the electron affinity of chlorine, I believe, which he had obtained by quite a different method than the one I had devised. My method depended on the current–voltage curves in a vacuum tube between a tungsten wire and a tantalum collecting electrode. I used the Langmuir relationships on the conductivity of such a system and its dependence on the mass of the charge carrier. By doing this first in a total vacuum, I knew that the carriers were electrons from the hot wire. By exposing that hot wire

George C. Glockler, Department of Chemistry, University of Minnesota, was my research supervisor from 1935 to 1939.

to various low pressures of the halogen(s), I was able to calculate the mass of the charge carriers. From that information and the equilibrium between the halogen molecules and halogen atoms that occurred at the hot wire, I could calculate the equilibrium between the halogen atoms and the electrons present at the surface of the hot wire. I translated this equilibrium calculation as a function of temperature into the energy of binding of the atom and the electron, that is, the electron affinity of halogen.

At the end of my second year I had to undergo what was called a "preliminary examination" before I could really begin to write my thesis. This examination was oral, as it is even today, but I cannot remember all the professors on my committee. However, one was a professor of analytical chemistry by the name of I. M. Kolthoff, and it was his prodding and questioning that caused me so much difficulty. He asked me to balance a dichromate oxidation of something, and I had great difficulty in doing this. In fact, I failed to do it! I don't recall whether they failed me and requested a second examination or simply asked me to leave the examination room so that the committee could deliberate. It was a traumatic experience. Nevertheless, I ultimately was admitted to candidacy. In later years I had occasion to meet Professor Kolthoff several times, and he reminded me of the event.

I. M. Kolthoff, Department of Chemistry, University of Minnesota.

Postdoctoral Education, 1935–1937

In the course of writing my thesis, I became familiar with the work of Professor Michael Polanyi of the University of Manchester in England. He originally was in Berlin, but at the time I began to know his reputation he had moved to England. Polanyi had begun the study of the nature of chemical reactions and developed the whole theory of transition states in chemical reactions at Manchester. Therefore, with Polanyi in the chemistry department, Manchester seemed like a very interesting place to work. The research seemed to be the type with which I would be pleased to be further associated.

I wrote to Polanyi in the hope that he might be able to help me obtain postdoctoral support to work in his laboratory. He was able to do this, fortunately, through the auspices of a grant that he had from the Rockefeller Foundation. Thus, at the end of my thesis work in Minneapolis I undertook to spend a year at the University of Manchester, England, working with Professor Polanyi. It turned out that I was a successor to Richard Ogg, another American, who became a professor of chemistry at Stanford University after he left England. This postdoctoral period began in the fall of 1935, and it was a most fortunate thing for my whole interest in chemistry that followed.

When I first met Michael Polanyi in Manchester in 1935, he was well into his second career. He had come to Great Britain in flight from Hitler in the early 1930s and had begun some particularly sophisticated studies. He was to conclude these studies within a relatively short time to begin another career, that of a political scientist—economist. There is no doubt in my mind that the experience I had with Polanyi was instrumental in opening my eyes to the advantages of an interdisciplinary approach to science, as he had the type of mind that was curious about all things, even those not directly involved with his current work. I kept in close touch with Polanyi over the years until his death in 1976 at the age of 85. It is particularly fitting that his son John won the Nobel Prize in chemistry (with Yuan Lee of Berkeley) in 1986 for his elegant work in atom—molecule reaction mechanisms, thus continuing the unique family career in chemistry that had been started by his father.

I feel that my experience with Professor Polanyi had a profound effect on my subsequent career, as did the immediately following experience with Gilbert N. Lewis in Berkeley. During the 2 years in Manchester I became aware of the freedom of thought that allowed me to undertake work in any area of science that seemed appropriate to the questions I was faced with. Thus, I was not limited to physical chemistry, organic chemistry, biochemistry, or biology, but encompassed them all to some degree.

Professor Polanyi had been interested in reactions of atoms because this kind of reaction was most easily susceptible to theoretical treatment. Upon my arrival I began an outgrowth

Michael Polanyi was professor of chemistry at the University of Manchester, England, at the time I did my postdoctoral work there (1935–1937).

of that work. First of all, I should point out that Polanyi had already done his work on sodium flames that led him to measure the rates of reaction of sodium atoms with alkyl halides and to formulate these reactions in terms of a three-center reaction. The sodium atom would approach the alkyl halide from either of two directions: from the halogen direction, pulling it off and leaving behind the carbonium ion, or from the other direction, in which the sodium atom would displace the halogen atom from the alkyl carbon. This latter turned out not to be the way in which the reaction went.

Thus began the whole idea of transition-state theory and the description of simple reactions in terms of a single reaction coordinate. In the case of the sodium atom and the alkyl halides, the coordinate was the line joining the carbon atom, the halogen atom, and the sodium atom. One could draw a potential curve showing the way in which the energy changes as a function of the sodium–halogen–carbon distances in various ways. It was this type of an idea that Polanyi wanted to apply to reactions of hydrogen atoms. The easiest way to get the hydrogen

Embarking upon postdoctoral studies, 1935.

atoms at that time was to allow hydrogen to come in contact with platinum metal, in which case the hydrogen molecule was supposed to dissociate into hydrogen atoms on the surface of the platinum, and thus a platinum–hydrogen bond would exist for which the reactions could be calculated.

It seemed clear enough that the platinum–hydrogen bond, whatever it was, was not a simple nonpolar bond but had some polarity. Presumably, the platinum was positive and the hydrogen negative (or it could have been the other way around). I was set the task of determining some of these properties by measuring the effect of the potential placed on the platinum electrode on the rates of some of the hydrogen atom reactions that would take place at that platinum electrode. Some of these reactions were the $H_2–D_2$ reaction and the para-hydrogen conversion, which was not dissociation but rather a magnetic effect. Another was the $D_2–H_2O$ exchange reaction. In any case, my work with Polanyi began by studying the effects of potential on a hydrogen reaction that took place at the platinum electrode.[2]

Polanyi had invented the method for measuring water density with an exquisite sensitivity. He had devised a micro method that was implemented by an extremely talented technician named Ralph Gilson. They made tiny glass floats (divers) that could contain in their shanks a few microliters of water and that had attached to the shanks a hollow glass bulb with a flat surface. The floating capacity of that bulb was determined by the pressure above it, which changed the volume of the bulb and thus its floating capacity. These bulbs were calibrated for different densities of water contained in the tiny chamber at the bottom of the float. We were able to measure the density of water on a few microliters to something like four or five decimal places, which was enough to determine the deuterium content to the accuracy that was required. The actual transfer of the water samples into the diver chambers was done by an especially trained young boy, who then transferred the divers to the water vessel under controlled pressure and measured the pressures at which it floated, neither rising nor sinking.

Most of our measurements at first depended on the change of deuterium content of the gas phase. Therefore another method, one not requiring combustion of the hydrogen to water, seemed appropriate. About that time Whytlaw-Gray in Leeds had devised a gas density balance that he was using to determine atomic weights (I think of the rare gases) very precisely. This seemed to Polanyi an appropriate method to be applied to the measurement of the hydrogen gas that we were using for the deuterium exchange reactions. The balance consisted of two glass bulbs mounted on the ends of a beam that, in turn, was supported on a bridge of quartz fibers. Here, again, the relative weight of each side was determined by the pressure and density of the gas in which we were interested, and proper calibration would allow us to determine the density of the gas to four or five decimal places. Because we were using only hydrogen and deuterium, we were able to determine the total deuterium content of any gas mixture. Neither of these methods, however, could be used to determine the amount of HD in a mixture of gases of hydrogen and deuterium. Another method had to be devised.

Polanyi did this by recognizing that the thermal conductivity of a gas mixture would depend not only on the number of hydrogen atoms and the number of deuterium atoms, but on how they were combined. In other words, thermal conductivity would depend upon how much HD was present in the mixture of H_2 and D_2 equilibrated to HD, which presumably required the activation of the hydrogen down to the hydrogen atoms. The HD measurement was accomplished by measuring the thermal conductivity of the gas mixture after it was calibrated in terms of the amount of hydrogen, deuterium, and HD in the mixture. This was sufficiently accurate for the task we had in hand. This method was calibrated with platinum as the catalyst.

I can very vividly remember rambling conversations that took place in Polanyi's office in Manchester toward the end of my first year, when I had been working exclusively on platinum—hydrogen activation systems. He pointed out that most biological oxidation—reductions were mediated via a porphyrin molecule such as heme, which was present in most of the biological oxidation—reduction catalysts, as well as chlorophyll, which is the photocatalyst in plants. If biological oxidation was a dehydrogenation and reduction—that is, a hydrogenation reaction that took place on porphyrins—then it should be possible to do a study on porphyrin molecules similar to the one that we had been doing on platinum, by using molecular hydrogen as one of the reagents. Of course, the natural biological oxidation–reductions, having porphyrins as prosthetic groups, took place on proteins. Polanyi speculated that the proteins were simply wires for moving electrons around to match the protons that had ultimately to be moved, and that the reactions took place on the porphyrin molecules themselves. This was, at least to my mind, a remarkable feat of association. Thus, it became important for me to find porphyrin molecules that might be useful in studying hydrogen activation in the same way we had been studying it on platinum.

We wanted to study the H_2-D_2 reaction that would produce HD in the gas phase on the surfaces of stable porphyrin-type molecules. This required a different method of measurement than the one that had been used for the exchange reactions $D_2(HD)H_2$. The exchange reaction, after all, could be measured by measuring the total amount of deuterium in the gas sample. This quantity could be determined by burning the gas (if it was hydrogen) or the organic compound (if it wasn't hydrogen) and measuring the density of the water so produced.

Our next task was to find a porphyrin redox catalyst model stable enough to be used in a gas-phase experiment at a variety of temperatures, including elevated temperatures. The ordinary porphyrins, even when most of the side groups were removed—that is, the naturally occurring porphyrins that could be obtained from hemoglobin or the catalase or any of the other porphyrin-containing oxido-reductases—were not stable enough for this purpose. About that time, Polanyi heard of the synthesis of a porphyrin analog in London by R. P. Linstead, one of the leading organic chemists of Great Britain. Linstead's discovery of the phthalocyanines, which really are tetraazaporphyrins, occurred as a result of an accident in the ICI factory that was making phthalonitrile. Apparently one of the glass-lined kettles, which were used to make phthalonitrile from phthalamide, cracked, and the whole reaction mixture turned a deep blue. Linstead, who was a consultant for ICI, was called in to help explain the chemical nature of the blue coloration. He was able to determine that the structure of the blue pigment so

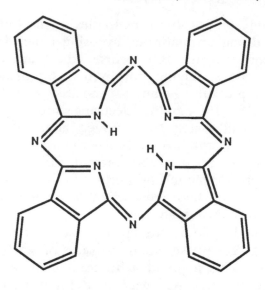

phthalocyanine

formed was the tetraazaporphyrin (now called phthalocyanine) and that its formation was catalyzed by the presence of the iron on the phthalonitrile.

The phthalocyanine turned out to be an extremely stable compound, and I was sent down by Polanyi to Linstead's laboratory at Imperial College in London to learn how to make and purify phthalocyanine. I spent a relatively easy few days in the laboratory because the synthesis of phthalocyanine was a simple procedure. All that was required was the heating of the phthalonitrile in the presence of a good tetracoordinating metal, particularly copper. However, once the copper enters the center of the phthalocyanine ring (or, in fact, the phthalocyanine is created around the copper ion), it is impossible to remove the copper. Even dissolving the copper phthalocyanine in concentrated sulfuric acid and then pouring that onto ice gives back the copper phthalocyanine. So we had to use other metals. The metal of choice turned out to be zinc, which was a very good template for the formation of phthalocyanine and could be removed by simple acid treatment to give the free tetraazaporphyrin, into which any metal could then be introduced.[3,4]

Having succeeded in preparing both metal-free and metal-containing phthalocyanines, we then undertook a study of the activation of molecular hydrogen in the gas phase by crystals of these materials. At first we thought we had indeed seen such an activation, but this later turned out to be a mistake, and the observation could never be repeated in a routine way. However, this kind of work was continued after I came to Berkeley as an instructor in 1937.

Before leaving this period, I would like to say a few words about the other students with whom I shared Polanyi's interest. The two principal ones were a pair of graduate students working on hydrogen activation problems in the laboratory adjacent to mine. One of them, E. C. Cockbain, worked for many years at the Malaysian Rubber Research Institute in London. The other, Dan Eley, became a professor of physical chemistry at the University of Nottingham. In fact, I recently (October 1986) saw him again when I gave the Holden Botany Lecture at Nottingham, as I did in 1958 when I received an honorary degree there. The group of students associated with Polanyi was a small one during the time I was in Manchester, and we each had our discussions with him separately. However, there was much exchange between the people in the basement laboratories at Manchester where we worked.

During the first year I was in Manchester, I really didn't take too much time for exploring England, although I had several trips to London to visit Linstead's laboratory and did see some of that city. After my return from a trip back to the United States during the first summer, I managed to find a little time to explore northern Britain in the form of a walking trip from Perth to Fort William in Scotland. I can't remember how many days it took, but it must have been at least a week or so. This trip was taken together with Jack Fidler, who was a Cambridge Ph.D. and postdoctoral associate in the Department of Biology at Manchester. I shared a house with him, and we each had a pair of rooms (bedroom upstairs and sitting room downstairs) parallel to each other. The house itself was situated on a street called Lady Barn Lane, about a mile south of the university on the way to the district of Didsbury. On several occasions Jack invited me to visit his family home, which was located on the River Thames just outside of Reading, and I spent several weekends there learning the behavior of a guest in a British country house.

Another trip I took from Manchester was during the Christmas holidays, when I went to Switzerland to ski. However, I cracked a leg bone and spent most of the 2-week holiday sitting on the hotel veranda and watching the other skiers go by!

The major focus of my stay in England was working in the laboratory, and the environs of Manchester at that time certainly weren't very conducive to outdoor activities. Manchester then, as now, was the heart of the midland industrial area. It was severely depressed economically during the time I was there.

At the time I was in graduate school, most of the organic chemistry that I was involved in (which wasn't very much, by the way) was again simply structural organic chemistry and some natural product chemistry. However, during my postdoctoral studies in 1935 I became conscious of the existence of the theoretical organic chemistry that was going on in the laboratories of Sir Robert Robinson at Oxford and Sir Christopher Ingold in London. This awareness became further ramified after I returned to the United States and began my career in Berkeley. At that time (1937) essentially all of the organic chemistry taught was natural product chemistry, with some synthetic studies and a few reaction mechanism studies in the style of Robinson and Ingold. When I began to discuss the theory of organic chemistry, both with Gilbert Lewis and later with G. E. K. Branch, the theories of Robinson and Ingold became much more visible, and the extension of Lewis's electronic theory, not only into structural chemistry but also into reaction mechanisms, became dominant.

Academic Appointment to the University of California, Berkeley, 1937–Present

The rest of my entire scientific career has been spent in Berkeley. In 1987, I celebrated my 50th year here. My appointment at Berkeley in 1937 was the result of a visit to Polanyi's laboratory by Professor Joel Hildebrand. Polanyi, who knew my situation and was aware of the fact that I had been in Manchester for a year and a half and was now ready to go back to the United States to seek an academic appointment, probably apprised Joel that I might be interested. I suppose Polanyi told him something of the quality of my commitment and suggested that he might consider me for an appointment at the University of California in Berkeley. I was introduced to Professor Hildebrand, and then he discussed with me the possibility of coming to Berkeley. When he returned to Berkeley he "convinced" Gilbert N. Lewis to hire me, which he did by mail. I was the first person who was not a graduate of the University of California to be hired in the Department of Chemistry since 1912.

Lewis assigned to me the first formal task on meeting me in the hallway outside his office, a place he usually spent most of the time. His office door was always open, and on it was a blackboard. Lewis would catch people as they walked through the halls to find out what they were doing and to give them whatever instructions he might have had in mind. The first instruction he gave me, before I had done any research work, was to undertake the arrangements for the weekly research conference held in the seminar room in Gilman Hall, just a few yards from his office. This responsibility was not mine alone. I shared it with Kenneth Pitzer, then a new instructor in the department who had just finished his degree with Professor Wendell Latimer in Berkeley that same year (1937). Together we would scurry about within the department to get faculty commitments for the research conference. This activity had two consequences: Pitzer and I got to know all of the faculty and graduate students and what they were doing, and the faculty got to know us.

Lewis wanted to discover my capabilities personally by working with me on various projects when I first arrived in Berkeley. In addition to the collaboration in the laboratory itself, Lewis and his wife, Mary, entertained the various new staff members frequently in their home. During these events I became acquainted not only with Gilbert Lewis himself, but also with his wife and three children; his two sons followed in his footsteps and became chemists. I can remember having the younger one, Ted, as a sophomore student in my beginning organic chemistry course. Another activity that the Lewises enjoyed was their participation in the drama section of the faculty club, an activity that showed the talents of the faculty and their wives or husbands as actors in plays of all types. (Incidentally, my wife and I were participants of this activity for many years and thoroughly enjoyed it.)

When I joined the staff at Berkeley, there were only three other organic chemists on the faculty of about 15: G. E. K. (Gerry) Branch (with whom I collaborated on a book on theoretical organic chemistry), T. Dale Stewart, and C. W. Porter. This group provided the nucleus for organic chemistry until after World War II. A more detailed explanation of the role that Gilbert Lewis played in the development of organic chemistry at Berkeley was discussed at a special symposium some years ago.[5]

In Berkeley the continuation of my hydrogen activation work took another turn. I began to study the activation of molecular hydrogen by using molecular complexes in solutions.

The kind of reactions studied were the H_2–D_2 reaction, the exchange reaction, and eventually homogeneous hydrogenation in solution with a variety of acceptors, such as quinone.[6,7] The first intimation of success came in the long study of the possible activation of hydrogen by copper compounds, starting out with cupric complexes of salicylaldehyde and salicylaldehyde ethylenediamine, which were square planar complexes. The copper complex, being paramagnetic, would catalyze the para–ortho hydrogen conversion, but that was not what we were seeking, because it did not involve dissociation of the hydrogen molecule. We were looking for H_2–D_2 reactions, hydrogen exchange reactions, and eventually hydrogenation reactions, by this system.

One of the experimental arrangements involved the placing of a solution of the copper complexes in contact with molecular hydrogen and then taking samples of the gaseous hydrogen in equilibrium with that solution for determination of the H_2–D_2 reaction. During one such experiment we had the

With Genevieve and Joel Hildebrand (right), who was responsible for bringing me to the University of California at Berkeley, celebrating at a reception in honor of my receiving the Priestley Medal, 1978.

copper salicylaldehyde dissolved in quinoline and kept at an elevated temperature in contact with molecular hydrogen. After a period of time the molecular hydrogen began to disappear, and the solution turned from green to red. We measured the kinetics of the rate of disappearance of the hydrogen and found it to be autocatalytic—that is, the rate starts out as zero, gradually increases until it becomes maximal, and then eventually it stops. This was traced to the reduction of the cupric copper to cuprous. After the copper was fully reduced to cuprous, the absorption of hydrogen ceased.

This development was interpreted as indicating that the cuprous complex dissolved in quinoline was truly the hydrogenation catalyst and that it then reduced the rest of the copper in an autocatalytic fashion. This interpretation led to the notion that a cuprous copper compound dissolved in quinoline might act as a homogeneous hydrogenation catalyst for some other material that could be reduced. This hypothesis indeed proved to be the case when quinone was added to the reaction, which started with cuprous compound. The reaction proceeded immediately and went linearly until the hydrogen acceptor was fully reduced. Simpler compounds were tried, and even cupric acetate would undergo such an autocatalytic reduction. This reaction produced cuprous acetate dissolved in quinoline, which also turned out to be a good homogeneous hydrogenation catalyst and could be used to reduce other materials. When cuprous acetate was the starting material, it was shown that the dissolved cuprous acetate was indeed the homogeneous hydrogenation catalyst that had been postulated.[7] Our last paper on homogeneous hydrogenation was published in 1956.[8]

Collaboration with Gilbert N. Lewis

As I indicated, this homogeneous catalytic hydrogenation was the first work undertaken during my first few years here in Berkeley, and it was a very exciting development. Until then (1938) all hydrogenation reactions had to have a solid surface catalyst for the activation of H_2. Such surfaces were either metals or metallic oxides. I myself did not pursue the work on homogeneous catalytic hydrogenation beyond my first few years at Berkeley. However, others, particularly Jack Halpern at the University of Chicago, have developed the idea in a more sophisticated way.[9]

I can remember doing this work on the third floor of Gilman Hall and being so excited by it as to run down to the first floor and tell Professor Gilbert N. Lewis, who was then dean of the college and chairman of the department, and who, as usual, was standing in the hall by his door blackboard. This led to an acquaintance with him that became more personal as time went by.

Lewis invited me to participate with him in writing a review on the color of organic substances, a project that took about a year.[10] It is perhaps worthwhile to review briefly the method by which this paper was written. It was ultimately published in *Chemical Reviews* in 1939.[11] Lewis would ask me to collect the data in the form of publications, either bound or unbound, and review books that contained information on the

29

color of organic substances. These books and journals were then laid out on tables in a room especially set aside for this purpose, open to the pages of relevance. Lewis would walk up and down these tables, looking at the data and making comments about it, and I would respond to those comments. After some weeks of this sort of activity, he said, "It is time to write." I had to get the paper and pencil, sit down, and start to take dictation from him. It turned out that although this was a relatively slow task, it was not difficult. We had spent so much time in our discussions of the subject that I knew pretty much what the words would be as they came forth, and I was usually ahead of him in writing them down.

It was a very enlightening experience, though, and one that brought me into contact with the electronic structure of organic molecules in a very detailed way. Lewis had written a chapter on color in his book *Atom and Molecule* many years earlier, in which the basic ideas were already expressed. He undertook in the *Chemical Reviews* paper to express the color of organic materials in terms of the mobility of their pi electron system.[11] He expressed that mobility in terms of the restoring force constant of a linear harmonic oscillator. It was relatively easy, at least qualitatively, to see how these electron oscillations varied with structure. Although it was not possible to calculate those restoring force constants from first principles, one could get some general empirical numbers that could be applied to new molecules. This is exactly what happened. It was a major introduction to the mobility of pi electron systems in organic molecules in such a way that the possible use of this information in other areas of organic chemistry became obvious to me. I then began to take an interest in using these concepts to interpret other aspects of organic molecular reactivity.

Another concept that arose during this time came from Lewis's recognition of the peculiarities of some of these organic molecules for re-emitting the excited light they had absorbed. Lewis recognized that one of the factors determining whether or not a molecule would fluoresce had to do with the rigidity of the structure—that is, the ease or difficulty with which the electronic excitation could be converted into rotational or vibrational energy. If this conversion was difficult, then the electronic excitation could be re-emitted as light. He then undertook to examine such molecules that did not re-emit their excitation energy as light, and to stiffen them up in such a way that they would. This approach led to freezing some of the triphenylmethane dyes, for example, in a glass and discovering not only that they would re-emit their exciting light, but also that, on some occasions the re-emission would not be of the exciting light itself, but of some longer wavelength and over a much longer period of time than ordinary fluorescence.

In order to understand this effect, Lewis conceived the notion that the first excited state of the molecule, having a high probability of being reached by light absorption, was of the same multiplicity as the ground state. In addition, it had an easy way to cross over thermally at some excited level into another state, in which the electron pairing changed so as to give rise to what he called the triplet state. Because of this thermal transition from singlet to triplet in the excited levels, the triplet state resulting from such a transition could not re-emit light very readily. To do so would involve a prohibited transition from excited triplet to ground-state singlet with light emission. This

Gilbert Newton Lewis at a vacuum line in his laboratory in Gilman Hall in the 1940s, about the time he was conducting research on color and the triplet state.

process gave rise to very long-lived excited triplets that appeared as phosphorescence. If this were so, Lewis reasoned, it should be possible to measure the magnetic susceptibility of these long-lived excited states, if indeed they were triplets.

To do this experiment we needed a balance, and for the balance we needed two quartz fibers, about 2 m long, from which to suspend the sample. Lewis would not let me turn the quartz rods in the glass-blowing torch. Lewis got the glass-blowing torch at one end of the 50-yard-long hall on the top floor of Gilman Hall. He took the quartz rod and stood over the torch, turning the rod (about 4 mm in diameter). When the rod was just the proper temperature, he handed me one end of the fiber, and I had to run down the hall with it while he held the other end. We kept repeating this performance until we got a couple of good fibers, which were then hung up inside glass tubes about 1 inch in diameter, joined by a similar horizontal tube at the bottom with ground joints at either end. The sample was suspended from these fibers in the horizontal tube, and the tip of the sample-carrying rod was watched through a microscope.

The next step in the experiment was to turn on the magnetic field and then turn on the light. My job was to tell Lewis which way and how much the sample moved. I said that it "jerked" the right way, but it came to rest going the wrong way; this went on for several days. We were illuminating the sample unsymmetrically at one end of the magnetic field, and the illuminated part of the sample should have come into the field; instead, it jerked and then went back the other way. Lewis didn't believe these observations. He then let me turn on the light and watched himself; the results were the same.

As you can imagine, this was a rather nervous time for me. The results were going the wrong way, and Lewis knew that the samples shouldn't behave in that fashion. He could not understand the results. Late one afternoon we were on the lower floor of Gilman Hall in the magnet room with the door open, and Professor William F. Giauque came by and asked what was the matter. Lewis explained what was happening. Giauque asked what gas we had in the glass tube, and Lewis replied that there was air in the tubes. Giauque said that air was not the proper gas because oxygen is paramagnetic. When we

William F. Giauque, Department of Chemistry, University of California at Berkeley. He received the Nobel Prize in chemistry in 1949 for his low-temperature work.

changed the air in the tubes to nitrogen, the experimental results were proper and as expected. The explanation of the first result was as follows: The sample became heated by the light, which warmed the oxygen, and the volume magnetic susceptibility became less so that the oxygen moved out of the field. The result was actually an oxygen wind blowing the sample in the wrong direction. Lewis said, "If that's the case, the same result should occur if I put a microscope slide with soot on it in the place of the sample." Everything did go properly. The next question was how much does it move in the right direction. From the cosine of the suspension displacement angle, we could determine the force acting on the sample. It turned out that the triplet state was indeed magnetic, and we closed up the show on that particular experiment in 1944.

In 1945 the first paper on paramagnetism of the phosphorescent state was published. It was the last paper I published with Gilbert N. Lewis.[12] In 1949 the last paper by Lewis on photomagnetism was published with Michael Kasha, Lewis's last graduate student.[13] A much more general consequence of this exercise was the growing interest in a variety of other properties of organic molecules that would be influenced by this delocalization phenomenon of pi electron systems.

Another consequence of the color study was the realization that one could classify, or subdivide, the absorption bands of such delocalized pi electron molecules in terms of the three dimensions in space. Most of the molecules, of course, were electronically only two dimensional (i.e., flat). Therefore, most of the molecules would have two main absorption bands, which Lewis called an X band and a Y band.[11] However, he predicted—and of course it was found—that if such flat molecules were allowed to aggregate in stacks, there should be a delocalization oscillation normal to the plane of the molecule itself in the direction of the stack. This he called a Z band. These molecules, of course, were found and described in detail by Scheibe[14] and others[15] for a series of cyanine dyes. These effects could be found even in simple polycyclic aromatic compounds when they were properly examined. Thus, the whole idea of the implication of electronic structure not only for color and spectra but also for quantitative interpretation of organic reaction mechanisms grew.

In general, as the previously described incidents might indicate, Lewis's scientific insights were highly intuitive in the first instance.[10] His mathematical prowess had some of the same qualities and was not of any very formal type. Nevertheless, having come to a conclusion as to what the scientific basis of a phenomenon was, he would then undertake to quantify it in a more normal mathematical way. Most of the time his mathematical discussion was relatively primitive. He would usually find a collaborator to whom he would explain the phenomenon, the fundamental basis for it, and a few simple mathematical ways of quantifying it. Then he would ask his collaborator to formalize it in a more conventional and detailed manner. This happened to me in the instance in which we were to calculate the expected horizontal movement of our magnetic balance when we were determining the paramagnetism of the triplet state, among other things. This intuitive approach to science was there in everything I saw Lewis do, and very quickly I came to appreciate the quality of that intuition. At first I was naive enough to raise fundamental questions when he did it, but as time went on I learned that Lewis always answered me rationally if he had the patience to do so. I learned later that others (Glenn Seaborg, for example) who worked with him in this way had exactly the same experience.

Collaboration with G. E. K. Branch

The Theory of Organic Chemistry

At that stage (about 1939) Lewis gave me the clear impression that I should work with Gerry Branch and collaborate with him in the preparation of a book on theoretical organic chemistry. Lewis told me that Branch had it all in his head, but he couldn't seem to get it written down. Therefore, it was my task, according to Lewis, to write down what was in Branch's head. This was not an order in any sense of the word. It was simply a suggestion from a colleague that this might be a fruitful collaboration, and so it turned out, as was usual with most of Lewis's suggestions. He must have said something to Branch as well, because he (Branch) invited me to come to work with him in that area.

In any case, we worked out an arrangement by which I would come to Branch's house for dinner once or twice a week. Branch would have written some material during the week, and I was supposed to write some more. We would rewrite and then plan the next week's task. That's how the book was written. I must have had 50 dinners at Branch's house (I was unmarried at the time). The book was finally published in 1941. Esther Branch, Gerry's wife and a former graduate student in the

G. E. K. Branch in 1936. (Reproduced from reference 5. Copyright 1984 American Chemical Society.)

Department of Chemistry, was heroically patient about the whole thing. She did all the cooking after her full day as a principal of a private school (Bentley School) in Berkeley.[5]

The Theory of Organic Chemistry by Branch and Calvin[16] was the first book on theoretical organic chemistry that had ever been written in the United States and contained quantum mechanical language (those were the chapters I wrote). The chapters written by Branch contained a detailed analysis of the effect of structure on the acidity of hundreds of organic (and inorganic) acids. Branch had already in his mind gone over the effect of structure on acid strength, as measured by pH and pK_a, and he used that as the foundation of the analysis of mesomeric (resonance, R_α) effects and inductive (polar, I_α) effects. These concepts are used to describe the effects of substituents and structure on acidity of any given series. The total effect is given in terms of a property of the substituent ($I_\alpha + R_\alpha$) and a property of the skeleton upon which the substituent is placed (A_α). This same type of explanation was used in Hammett's book, which appeared about the same time as ours, much to our surprise.[17] One constant had to do with the effect of substituent (σ) and the other constant with the effect of structure (ρ) upon which the substituent was placed.

The publication of *The Theory of Organic Chemistry*[16] in 1941 by Branch and me was the beginning of theoretical organic chemistry in the United States. Our book, in effect, organized all of organic chemistry in terms of electron theory. Physical organic chemistry at Berkeley and in the United States depends upon Gilbert Lewis's initial stimulus in the concept of the electron-pair bond and Branch's evolution and development of that concept and stimulus not only for structure but for reactivity and mechanism as well. A summary of the influence of Lewis and Branch on the development of theoretical organic chemistry at Berkeley was given at an American Chemical Society National Meeting in Las Vegas in 1982 and later published in the *Journal of Chemical Education*.[5]

There is no question in my mind that Gerry Branch was one of the greatest theoretical organic chemists of his day. The fact that Lewis had recognized this and had felt because of experiences, particularly with Polanyi, that we could create a useful book, was an important step in my development at Berkeley. The publication of this book made Berkeley one of the foremost centers in the United States for theoretical organic chemistry.

Collaboration with Biological Departments at Berkeley

About this time, 1940 or so, I was introduced to molecular genetics by virtue of becoming acquainted with Professor Richard Goldschmidt of the genetics department. He had an assistant by the name of M. Kodani who had done some work on the effect of acid and base, particularly alkali, on the structure of chromosomes as viewed on a microscope slide. The banded structure of the salivary gland chromosome was clearly visible, and when it was treated with alkali on the slide, profound rearrangements occurred. This mild treatment seemed to me to require that the structure of the chromosome was maintained by a collection of hydrogen bonds, particularly of the protein that surrounded the nucleic acid of the chromosomes. The banded structure seemed to me to have been set up, or established, by the stacking of the nucleic acid bases, and a short paper was published to this effect in 1940.[18] The idea of the stacking of bases in the nucleic acid derived from our previous work on the stacking of dyestuffs and their X, Y, and Z absorption bands.[10] It seemed to me that the same stacking phenomenon was occurring in the nucleic acid in the chromosomes.[19] This was the first clue to what turned out to be the structure not only of the protein, which had already been suggested as a hydrogen-bonded structure by

Linus Pauling,[20] but also of the stacking of the bases in the nucleic acids. The stacking idea came from our own concepts derived from the stacking of organic dyestuffs, first seen by Scheibe.[14]

Later, in the early 1950s, I had a visit from James Watson in my office in the old Chemistry Building. He wanted to know what I thought about the nature of the tautomerism in the purine and pyrimidine bases. He didn't tell me why he wanted to know this, and I didn't ask him. However, I sat down with him and drew pictures of each of the bases and tried to estimate which of the tautomers would be the most stable in each case. This estimate was based on the much earlier work I had done with Branch on tautomeric equilibrium and problems with him that were described in *The Theory of Organic Chemistry*.[16] This is probably the reason Watson came to see me.

The War Years

In 1942 our thoughts turned to what we might be able to contribute to the war effort. I had already noticed the papers of Pfeiffer[21] and Tsumaki[22] on the ability of cobalt coordination compounds of a specific type to reversibly bind oxygen. This consciousness arose out of the continuing and growing interest in coordination chemistry that was generated by our work on the porphyrins and the metalloporphyrins, which had begun with Polanyi in Manchester, and also out of the knowledge that reversible oxygen binding by hemoglobin was a property associated with the porphyrin. When Tsumaki's work appeared in 1938, I immediately became aware of it and started some work with these cobalt compounds in the basement of the old Chemistry Building.

About that time I visited Wendell M. Latimer, who was then dean of the College of Chemistry, to see if there was anything I could do to assist in the war effort. He was our principal connection to Washington, D.C., largely through the National Defense Research Committee (NDRC), of which James Bryant Conant was chairman. I told Latimer what I knew about the ability of cobalt salicylaldehyde ethylenediamine compounds in the solid state to reversibly bind molecular oxygen and the possibility of using this phenomenon as a means of generating molecular oxygen on board ship without the necessity of carrying compressed gases. This intrigued him very much. He

40

Wendell M. Latimer, dean of the College of Chemistry, University of California at Berkeley, during World War II.

arranged for me to go to Washington to meet Professor James Bryant Conant, tell him about this idea, and see if something could be done to get support. I did this, and returned with encouragement and some support to develop some of these compounds with a view toward building an oxygen-generating machine that could be mounted on either a destroyer or a submarine in such a way that the amount of compressed gases required would be greatly reduced.

Our more general concern about the basic structural features that determined the binding of metal ions by chelating groups was well underway. We thus spent some time developing the cobalt salicylaldehyde ethylenediamine compounds, and their varieties, as reversible oxygen carriers. This work was summarized in a series of papers that appeared after the war in 1946 and 1947, in which the properties of the chelate compounds were described and the methods of synthesis were outlined.[23] We eventually developed the trifluoromethyl-substituted salicylaldehyde as the most stable starting material for these cobalt

Professor James Bryant Conant of Harvard, chairman of the National Defense Research Committee.

compounds. It could then be cycled many, many times, absorbing oxygen from the air and desorbing it as pure oxygen in order to provide a good source for both breathing and welding oxygen, for example.

The basic principle that emerged from this work was that the salicylaldehyde ethylenediamine cobalt compounds bound oxygen in the ratio of two cobalt atoms to one oxygen molecule, as follows:

$$2CoL^1 + O_2 \rightleftarrows (CoL^1)_2O_2 \qquad (1)$$

We also made compounds in which the ethylenediamine component was replaced by 1,5,9-triazanonane. This new compound had an extra nitrogen atom in it, the center one of the nonane, which could reach around and fill a fifth coordination position of the cobalt atom. In solution this cobalt compound showed a simple equilibrium with oxygen of one cobalt per oxygen molecule and, of course, without a phase transition in solution. It was, however, not very useful for our oxygen-producing purposes:

$$CoL^2 + O_2 \rightleftarrows CoL^2 \cdot O_2 \qquad (2)$$

A further general conclusion of our studies on the solid-state oxygen equilibrium was that it involved a phase transition such that the oxygen partial pressure over the crystals was constant for all degrees of oxygenation of the crystal except at each end. The initial saturation was reached very quickly, followed by a constant equilibrium pressure until the last 5%, at which time the equilibrium pressure rose. The instability of the crystal was, of course, a major problem that had to be resolved. It turned out that the principal point of attack on the cobalt salicylaldehyde ethylenediamine molecule in a crystal during cycling was the position ortho to the phenolic group in the salicylaldehyde. We made a large series of substitutions at this point and finally decided that the substitution of the trifluoromethyl (CF_3) group at that position was the best stabilizing influence we could produce.

In fact, the work went so far as to have us build a small heat exchanger as a demonstration unit. This consisted of brass cylinders, about 6 inches in diameter, surrounding each of two headers into which about 20 1.5-cm brass tubes had been mounted. The cobalt compound had been pelletized and put into the inside of the brass tubes. Around the outside we were able to pass cold water or steam, depending on what gas was coming in or out of the gas tubes. Thus, when air blew through the heat exchanger, oxygen would be absorbed by the cobalt compound and heat would be evolved; we would have to pass cold water around the outside of the heat exchanger tubes to remove the heat. After the cobalt was saturated with oxygen, we could shift the connection such that the gas would be given off into a container when we passed steam around the outside

The old Chemistry Building, University of California at Berkeley, 1890s, was razed in 1963 to make room for Hildebrand Hall. Its cupola was saved and is displayed in Giauque Plaza, between Gilman and Hildebrand Halls. I maintained an office there from 1945 to 1963. Three tall windows (right) mark the location of the office. When I moved to the Laboratory of Chemical Biodynamics, I took both the fireplace and the beautiful wooden cabinets with their glass doors. In 1980, they again traveled with me to my present office in Latimer Hall.

of the tubes. Thus, we had a cyclic oxygen generator that could generate pure oxygen from ordinary air.

I can remember taking this entire unit to Boston, aboard a series of DC-3s. Remember, this was a few years after 1940, and traveling was very difficult. I made the trip to the Massachusetts Institute of Technology with the apparatus in the hold of an airplane, and I must have changed planes half a dozen times: San Francisco to Los Angeles, Los Angeles to Dallas, Dallas to somewhere in Oklahoma, thence to Memphis and to Knoxville (Tennessee), on to Washington, and finally to Boston. At each transfer point the device would have to be moved from one DC-3 to another and I had to stand at the entrance to the plane to see that it was indeed moved as planned. This was a prototype unit that might be tried on board ship. My recollection of the events after that is not very clear. I believe the Arthur D. Little Company made a device, but by that time the liquid air method of generating pure oxygen had superseded any alternative method and was the one of choice. The method I developed never did get much beyond the prototype that I built.

Having done this work, our interest in the basic principles of chelation was greatly expanded. Toward the end of the war, about 1943 or 1944 I should think, Glenn Seaborg had already begun the work on plutonium extraction. In fact, the big plants had been built at Richland and Oak Ridge for the isolation and purification of plutonium and the decontamination of uranium. Nevertheless, I became involved in, first, the problem of decontaminating—that is, purifying—irradiated uranium of fission products and, second, the problem of isolating or purifying plutonium. Of course, I took this up in terms of a chelation and extraction problem rather than a precipitation problem. Seaborg had solved the same problem in terms of a precipitation method that involved bismuth phosphate; this was the first method used. In fact, I think the first large isolation plant to use the bismuth phosphate method was already under construction. By the time I got into this work, I had only to improve the method and change it from a precipitation method to a solvent-extraction method—at least, that was my charge.

To do this I knew that I would have to build a coordination, or chelation, compound that would be able to bind selectively the 3+ and 4+ elements—that is, the rare earth elements—while leaving behind the uranyl(2+) ion in acid solution. I did this by designing a β-diketone (thenoyltrifluoroacetone, TTA) that was sufficiently acid to succeed in the process. This was done by placing a CF_3 group on one side of acetylacetone and a thiophene group on the other. The CF_3 and thenoyl groups increased the acidity of the enol to such an extent that it did indeed bind 3+ and 4+ elements in nitric acid solution, and the thenoyl group retained benzene solubility (hydrophobicity). The synthetic effort was accomplished by several postdoctoral people who were involved in this project. I personally did no "hands on" work, but I directed the efforts of others. The synthesis was a simple one:

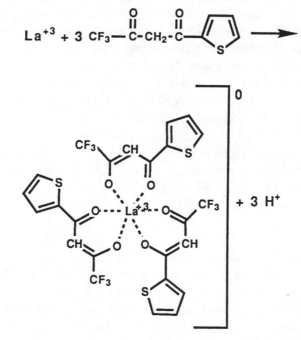

These coordination compounds, no longer charged because of their binding, could be extracted in organic solvent such as benzene from the aqueous acid.

The method was indeed a successful one and could be adjusted to extract one element after another by adjusting the acidity and the salt concentration of the solutions. By this time the plants in Richland had been built and the plutonium was already produced. Our solvent-extraction method was not introduced in the United States during the war. It was, however, examined both at Chalk River in Canada and at the British Atomic Energy Research Establishment at Harwell. In 1950 our last paper that had been done during the war years on the chelation extraction of plutonium and fission products was published.[24] Thus, our work on solvent extraction was completed.

As a consequence of these two major activities in chelation chemistry for practical purposes, our general interest in the nature of chelation and the way in which groups were bound, particularly to transition elements, was stimulated. Shortly after the war Professor Arthur E. Martell, at that time a member of the faculty of Clark University in Worcester, Massachusetts, spent a sabbatical year with me. His interest in chelation chemistry stimulated me to consider the possibility of writing a book with him on the chelate compounds. This we undertook, and after some time the book (*The Chemistry of Metal Chelate Compounds*)[25] was indeed completed and published in 1952.

Arthur E. Martell (now at Texas A&M), who co-authored The Chemistry of Metal Chelate Compounds *with me.*

I suppose the single most important generalization that came out of this work, aside from the enormous collection of stability constants for chelation as well as thermodynamic data (entropy and enthalpy changes) for such binding, was the idea that the principal driving force in chelation, as opposed to ordinary single-ligand coordination, was the fact that there was generally a very large entropy increase when the metal was bound by a chelate group. It seemed that the water released from the hydrated metal ion was a principal source of this increase in entropy, something that happened in a single step for two or more ligands, as opposed to the addition of a single ligand at that point.

Collaboration with Genevieve Calvin on Rh Work

Five years after I came to Berkeley I married Genevieve Jemtegaard, a juvenile probation officer. We met through mutual friends and married within a month. This was in 1942, at a time when her profession was almost completely male dominated. A graduate of the University of California at Berkeley and an outstanding debater, she was well qualified by her university training to undertake the profession she chose. In fact, her friends on the debating team provided the introduction. These friends have become practicing attorneys in Los Angeles, primarily involved in patent law, and we have been in constant contact ever since.

After the death of our first child due to Rh incompatibility, Genevieve became my collaborator in an effort to determine the chemical factors that caused this difficulty. Again, I undertook an interdisciplinary effort with a bacteriologist—microbiologist and a medical doctor.[26] Experiments were performed in the old Chemistry Building, using a special ultracentrifuge funded by the Rockefeller Foundation for this particular work. We made an effort to isolate the haptenic material (which we named elinin for our daughter, Elin) from Rh-positive cells to be used as a blocking agent during the course of an incompatible pregnancy. This did not turn out to be useful, and we were never able to get the pure haptenic material, but it did give us a chance to become acquainted with the nature of the immune reaction.

The experiments resulted in some definition of the chemical structure of this Rh incompatibility factor, elinin, and various serological tests were performed to discover whether or not this avenue of research should be continued. Genevieve's name appears on the publications as a co-author because she spent a great deal of time actually in the laboratory working with the antigenic material, helping to isolate it from over-aged blood collected from the blood banks.[27] This was her first chemical laboratory experience but not her last by any means, as will be seen in a later part of this book.

Establishment of the Bio-Organic Group of the Radiation Laboratory

The work on plutonium extraction and uranium decontamination that originated at the suggestion of Professor Wendell M. Latimer brought me into contact with Professor Ernest O. Lawrence, who was the director of the Radiation Laboratory, now called Lawrence Berkeley Laboratory, or LBL, in Berkeley that played a major role in the development of the initial production of plutonium. I met him on several occasions in Berkeley, but even more frequently on our periodic meetings in the Metallurgical Laboratory at the University of Chicago where all the workers in the very secret plutonium project were brought. At that time the project that ultimately was called the Manhattan Project was known as the Metallurgical Project for security reasons. The connection with Ernest Lawrence developed further in Berkeley. I got to know him personally because of the fact that I ate lunch at the so-called "physics" table at the Men's Faculty Club, where Ernest Lawrence also came. Because of this I became more personally acquainted with him than might otherwise have been the case.

Toward the end of 1945, particularly after the end of the war with Japan, Ernest Lawrence indicated to me that it was time to do something "useful", as he said, and thus expand our interests beyond the uranium—plutonium fission product extraction procedures. This was on one of the occasions when we walked back from lunch at the Faculty Club to our respective buildings, Ernest to LeConte Hall (the physics building) and I to Gilman Hall (the chemistry building).

By that time there was available a small amount of carbon-14 that had accumulated in the nitrogen-containing materials that had been stacked around the cyclotron for neutron-shielding purposes. This ammonium nitrate turned out to be rich in carbon-14, which could be recovered by adding a little carrier bicarbonate to the solution, and precipitating barium carbonate, which carried along the carbon-14 that had been generated by the neutron reactions on nitrogen. We thus had a relatively large supply—in retrospect, we had the world's supply—of carbon-14. Lawrence suggested that we undertake to use this material both for a study of organic reaction mechanisms and for the possible construction of radioactive compounds that could find their way to specific malignancy sites and thus, depending on their radiation, destroy those malignancy sites.

This work with the organic reaction mechanism studies and the synthesis of ^{14}C-labeled compounds of interest was begun in the Donner Laboratory. The Donner Laboratory had been established just prior to World War II by William Donner on behalf of Ernest Lawrence's younger brother, John H. Lawrence, who was a physician interested in the treatment of cancer, particularly with high-energy particles from the cyclo-

Professor Ernest Orlando Lawrence was director of the Radiation Laboratory, since renamed the Lawrence Berkeley Laboratory in his honor. (Photo courtesy of the University of California at Berkeley.)

tron. Ernest's concern to help his brother, John, led him to make the proposal I have just described. The space made available for our synthetic work and reaction mechanism studies was one-half of the third floor of Donner Laboratory, and we remained there for 18 years until the entire Bio-Organic Group came together as one entity in the Laboratory of Chemical Biodynamics.

At the same time, Lawrence made available parts of the old Radiation Laboratory, where the old 37-inch cyclotron was still sitting but where there was vacant laboratory space that we could use. We moved in and began our work on tracing the path of carbon in photosynthesis. This effort began roughly in 1946.

Just before the war, Sam Ruben and Martin Kamen had discovered the radioactive isotopes of carbon, particularly carbon-14. They had done a few experiments with carbon-11, which was readily available from the cyclotron but which had a half-life of only 20 min. They undertook to examine what the first product of green plant photosynthesis would be using ^{11}C because the supply of ^{14}C was minuscule[28]. What Ruben and Kamen did establish was that the radioactive carbon from CO_2 appeared first in a carboxyl group, and that was as far as they got. During the war Martin Kamen was taken up with uranium separation work and Sam Ruben undertook work with war gases, particularly phosgene; it was an accident with phosgene that killed Ruben. Thus, both of the originators of the work on photosynthesis using radioactive isotopes of carbon were separated from it.

I inherited from Ruben a small vial of barium ^{14}C-carbonate that had been isolated from the ammonium nitrate tanks around the cyclotron. With this carbon I began to carry out the work that Ruben and Kamen had started, a natural step for me because I had long been concerned with the oxidation–reduction properties of porphyrins, including heme and chlorophyll. So, my interest in plant redox chemistry had already been established. The problem of determining the path of carbon in photosynthesis was pretty well defined. We had known for more than 100 years that green plants absorbed CO_2 from the atmosphere and reduced it to a variety of sugars with a concomitant evolution of oxygen. The question was, What were the reactions involved in getting from CO_2 to sugars?

The old Radiation Laboratory (ORL), University of California at Berkeley. Constructed about 1885, it was replaced by Latimer Hall in 1958. To the right was Crocker Lab, which housed the 60-inch cyclotron, successor to the original 37-inch cyclotron from the ORL, and site of many important experiments during World War II. At the far end of the alley was the Anthropology Museum, one of the red brick buildings from the earliest days of the university. All have been replaced by more modern buildings.

A rather extensive program in reaction mechanisms was undertaken in the Donner Laboratory. We not only had carbon-14 to trace organic reaction mechanisms, but we also had deuterium. It was during that period that we found the isotope mass effect on chemical and biochemical reactions. In the case of carbon-14, the mass effect was first noticed in the decomposition of malonic acid. We saw it again when we began work on the path of carbon in photosynthesis. In the old Radiation Laboratory we had constructed some equipment in which we used barley leaves in a closed system and circulated the CO_2 containing carbon-14. We had two measuring devices; one measured the CO_2 by an infrared measurement of essentially the carbon-12, and the other detection method measured the radioactivity— that is, the carbon-14. We found that as the plant used up the carbon dioxide, the specific activity of the remaining carbon dioxide rose; that is, the amount of ^{14}C per mole of total CO_2 remaining in the residual carbon dioxide after the plant had used up 90% of it was considerably higher than the specific activity of the carbon-14 in the CO_2 with which we began the experiment. This finding indicated that the plants

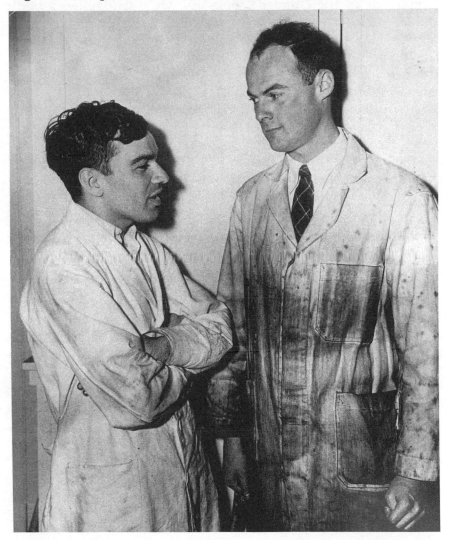

Martin D. Kamen (left) and Samuel Ruben discovered the radioisotope carbon-14 in the old Rad Lab in 1940. This photograph hangs in the Lawrence Hall of Science, University of California at Berkeley.

selectively used carbon-12 somewhat faster than they used carbon-14, and thus we saw the isotope effect in photosynthesis for the first time.[29] An effort was made by the local Radiation Laboratory lawyers to derive a patent on isotope separation from this observation, but it didn't seem practical to me.

As a result of our work on the synthesis of ^{14}C-labeled compounds in the Donner Laboratory, we put together a book entitled *Isotopic Carbon*[30]. The authors were Charles Heidelberger, James C. Reid, Bert M. Tolbert, Peter E. Yankwich, and I. This book, published in 1949, described the synthesis of a large group of compounds and became the synthetic "bible" for the early workers with radiocarbon.

In the middle 1950s we found that some of the ^{14}C-labeled compounds that had been prepared some years before were decomposing. It turned out to be a result of self-irradiation from the the carbon-14 in the compounds. The case that was particularly spectacular was the decomposition of choline chloride, which turned out to be a chain reaction that gave rise to a very large number of decomposed molecules for each electron given out by the carbon-14.[31] As a consequence of this observation, we went on in 1956 to try and label organic compounds (mostly aromatic compounds) by bombardment with accelerated carbon-14 ions.[32]

The decomposition that accompanied such an irradiation required the purification of the material after it had been irradiated. Furthermore, the labeling was nonspecific. Nevertheless, it was a relatively easy way to produce labeled materials, a process that was studied for carbon-14 and tritium.

The Path of Carbon in Photosynthesis

In 1945, when it became apparent that carbon-14 would be available cheaply and in large amounts by virtue of the nuclear reactors, and with the encouragement and support of Ernest Lawrence, we began to study that part of the energy-converting reactions of photosynthesis represented by the carbon reduction sequence, making use of carbon-14 as the principal tool. This work encompassed the period 1946–1956, during which time all of the individual steps in the carbon reduction cycle were put together. There were 23 publications entitled "The Path of Carbon in Photosynthesis" and two books that I coauthored with James A. Bassham.[33,34] The significance of the elucidation of the multisteps in the path of carbon was recognized by the award of the Nobel Prize in 1961.[35]

One of the main difficulties in the investigation of photosynthesis—in which the machinery that converts the CO_2 to carbohydrates and the substrate upon which it operates are made with the same atoms (carbon and its near relatives)—is that ordinary analytical methods do not permit us to distinguish easily between the machinery of the reaction and the substrates. However, the discovery of the long-lived isotope of carbon (carbon-14) by Sam Ruben and Martin Kamen just before World War II provided the ideal tool for tracing the path of CO_2 on its way to carbohydrate.

Design of the Experiment. The principle was simple. We knew that the CO_2 that enters the plant ultimately appears in the plant materials, but primarily, and in the first instance, in carbohydrates. Our purpose was to shorten the exposure time to such an extent that we might be able to discern the path of carbon from CO_2 to carbohydrate as the radioactivity enters with the CO_2 and passes through the successive compounds.

For this purpose we used reproducible biological material, first as barley shoots and later in the form of the unicellular green alga, *Chlorella pyrenoidosa*. We developed methods of growing the algae in a reproducible fashion in intermittent and continuous cultures. The algae were initially exposed to the radioactive CO_2 in a simple apparatus called a "lollipop", which contained an algal suspension with normal CO_2. Some ^{14}C-labeled CO_2 was injected into the stream of nonradioactive CO_2 for a suitable period, ranging from seconds to many minutes. At the end of the preselected time, the algae were killed by various methods. The first paper on "The Path of Carbon in Photosynthesis" was published in 1948.

"Lollipop" glass apparatus for photosynthesis experiments using $^{14}CO_2$.

In the old Radiation Laboratory with algae culture flasks.

Early Analytical Methods. In the early work the classical methods of organic chemistry were applied to the identification procedures, but these were much too slow and would require extremely large amounts of plant material to provide identification of specific labeled compounds. Here we were able to call upon our experience during the war years when we had used ion-exchange columns for the separation of plutonium and other radioactive elements. We used both anion and cation-exchange

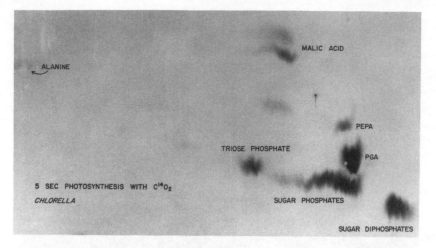

Two-dimensional paper chromatogram of a 5-second experiment with Chlorella pyrenoidosa *algae. The radioactivity has concentrated predominantly in one spot (later identified as phosphoglyceric acid) in such a short exposure.*

columns and soon discovered that the compounds that became radioactive in the shorter time exposures were anionic in character.

Because of the difficulty we found in eluting the principal early radioactive components from anion-exchange resins, it became apparent that this radioactive material was strongly acid and very likely had more than one negative point of attachment to bind to the cationic sites on the resin. A more detailed analysis of the precise conditions required to elute the material off the ion-exchange columns suggested phosphoglyceric acid as a possibility because of its already-known role in glycolysis.

Paper Chromatography. About this time, Consden, Gordon and Martin[36] in England had developed their method of partition paper chromatography, which was particularly well adapted for analyses of the type we were attempting because of the sensitivity of the colorimetric detection method and the ease of finding the radioactive compound on the paper by exposing it to an X-ray film. We turned to paper chromatography as our principal analytical tool. It was clear that the coordinates of a particular radioactive spot on a particular two-dimensional paper chromatogram could be interpreted in terms of chemical structure in a general way.

Our procedure was to find other properties of the material, such as fluorescence or ultraviolet absorption, and then elute the radioactive material from the paper, as defined by the black area of the X-ray film upon which the paper had been placed, and to perform chemical operations on the eluted

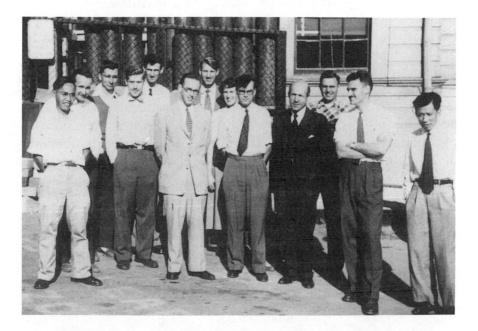

With my staff, outside the old Radiation Laboratory, 1954. Left to right: Ning Pon, then a graduate student, and now working in the Calvin Laboratory; James Bassham, who worked with me from 1946 until 1985; Jean Bourdon, Eastman Kodak, Paris; R. Clinton Fuller and Rodney Quayle; Hans Kornberg, Department of Biochemistry, Cambridge University, England; Hans Grisebach, University of Freiburg, Germany; Alice Holtham, then the departmental secretary; E. Malcolm Thain; me; Paul Hayes, Laboratory Manager, who worked for me from 1951 until his retirement in 1984; Jacques Mayaudon, postdoctorate student from Belgium; and Kazuo Shibata, Institute of Physical–Chemical Research, Japan.

material. The resulting chemicals were rechromatographed to gain some clue as to what the particular chemical procedure might have accomplished. The final identification was achieved by cochromatography of the tracer amount of unknown material with carrier amounts of the authentic suspected compound. Then a suitable chemical test, to which the added authentic material alone could respond, was performed on the paper. If this response coincided exactly with the radioactivity on the paper in all its structural details, we could be confident that the radioactive compound and the carrier were identical.

PGA as the First Product of Photosynthesis. It was clear that in only 30 seconds the carbon had passed into a wide variety of compounds and that it would be necessary to shorten the exposure time to get some clue as to the compounds into which CO_2 is incorporated earliest. We did this in a systematic way, and it became quite clearly apparent that a single compound, phosphoglyceric acid (PGA), dominated the picture in fractions of a second, amounting to 80–90% of the total fixed radioactive carbon.

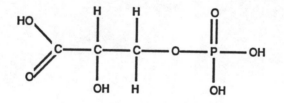

Phosphoglyceric acid

In trying to determine the origin of the PGA itself, we were led by what appeared to be an obvious kind of arithmetic to seek a compound made of two carbon atoms as a possible acceptor for the radioactive CO_2 to produce the carboxyl-labeled three-carbon compound, PGA. A good many other compounds were identified in the course of this search, in particular a five-carbon sugar (ribulose as its mono- and diphosphate) and a seven-carbon sugar (sedoheptulose as its mono- and diphosphate).

The time relationship of the trioses and hexoses to PGA seemed clear, but the sequential relationship of the five- and seven-carbon sugars was not easily determined. The distribution of radioactivity in the pentose and heptose was next determined. It turned out that the number 3 carbon atom of the ribulose is the first to be labeled, followed by carbons 1 and 2 and finally by carbon atoms 4 and 5. In the sedoheptulose the center three carbon atoms were first, followed by carbon atoms 1 and 2 and then 6 and 7 (Chart I).

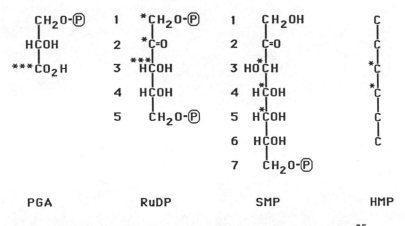

Chart I. Distribution of radioactive carbon in certain sugars.[35]

Identification of the Two-Carbon Acceptor. The next problem was to determine the source of the pentose with its peculiar and unsymmetric labeling pattern. It turned out that the only remaining alternative for the construction of the pentose was the combination of a C_3 with a C_2 fragment. The ribulose labeling scheme indicated that, following carbon atom 3, the next label appeared about equally in carbons atoms 1 and 2. When we realized that the ribulose we were degrading, which we obtained from the ribulose diphosphate, actually had its origin in two different reactions, it then became possible to devise a scheme for its origin, as shown in the lower left corner of Scheme I.

Taking the two-carbon fragment from the top number 1 and 2 carbon atoms of the sedoheptulose and adding it to glyceraldehyde (triose) could make another five-carbon compound. Taken together with the five-carbon compound remaining from the sedoheptulose, this compound would produce the labeling scheme finally observed in ribulose diphosphate (Scheme II).

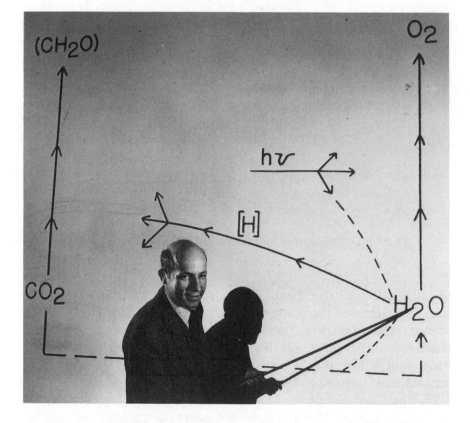

Discussing a simplified version of the carbon cycle.

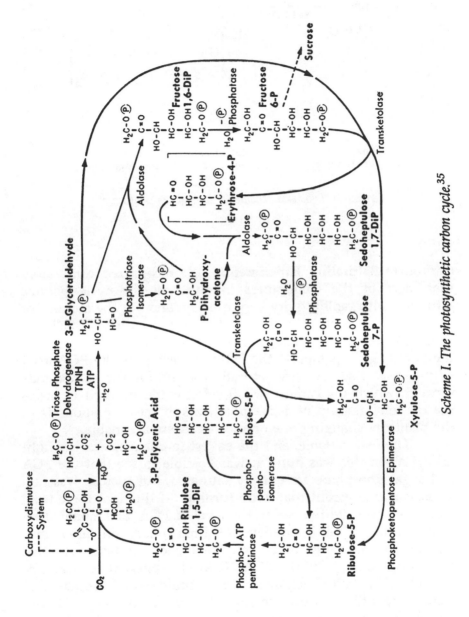

Scheme I. The photosynthetic carbon cycle.[35]

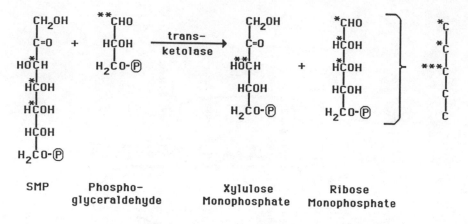

SMP   Phospho-   Xylulose   Ribose
        glyceraldehyde   Monophosphate   Monophosphate

Scheme II. Proposed scheme for labeling of pentose.[35]

Carbon-14 Saturation Experiments. We recognized quite early that most of the compounds involved in the cycle became saturated with radioactivity very quickly, and yet the amount of these materials present in the plant at any one time is small and does not change. This suggested a method for discovering how the light might operate on the PGA and also how the PGA itself might arise. Radioactivity saturation levels for the compounds through which the carbon was flowing could be used to measure the total amount of active pool sizes of these compounds in the plants by changing one or another external variable.

The first variable, and the easiest to change, was the light itself. When this was done, it was possible to see that the PGA and sugar phosphate are quickly saturated, but sucrose was not. Thus, it was apparent that upon turning off the light there is an immediate and sudden rise in the level of PGA, accompanied by a corresponding fall in the level of diphosphate, primarily ribulose diphosphate (Figure 1). We then had our first definitive clue as to the origin of the PGA. It would appear that it came as a result of a dark reaction between ribulose diphosphate and CO_2. It was thus possible to formulate the cyclic system driven by high-energy compounds produced in the light acting upon PGA.

The Photosynthetic Carbon Reduction Cycle. All of the individual steps in the path of carbon in photosynthesis were put together in a sequence, now called the Calvin cycle, shown in Scheme I. The efforts of approximately 10 graduate students, 25 postdoctoral visitors, and the permanent support staff were required to complete this monumental task, which took us approximately 10 years. Support for this effort came from the U.S. Atomic Energy Commission (now the Department of Energy); it still continues. The significance of the elucidation of the multisteps in the path of carbon was recognized by the award of the Nobel Prize in chemistry in 1961.

Before leaving this topic I would like to describe a moment (and, curiously enough, it *was* a moment) when the recognition of one of the basic facets in the photosynthetic car-

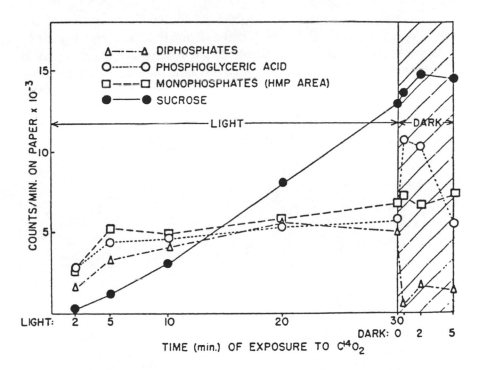

Figure 1. Effect of light and dark on photosynthetic reactions. (Reproduced with permission from reference 35. Copyright 1962.)

bon dioxide reduction cycle occurred.[37] One day I was sitting in the car while my wife was on an errand. For some months I had had some basic information from the laboratory that was incompatible with everything that, until then, I knew about the cycle. I was waiting, sitting at the wheel of the car, probably parked in the red zone, when the recognition of the missing compound occurred. It occurred just like that—quite suddenly—and suddenly, also in a matter of seconds, the cyclic character of the path of carbon became apparent to me. But the original recognition of phosphoglyceric acid, and how it got there, and how the CO_2 acceptor might be regenerated, all occurred within a matter of 30 seconds. So, there is such a thing as inspiration, I suppose, but one has to be ready for it. I don't know what made me ready at that moment, except that I didn't have anything else to do but sit and wait. And perhaps that in itself has some moral.

Perhaps the sudden insight into the cyclic character of the path of carbon might have resulted from the fact that the right answer came about as the result of an instinctive creative trick. That trick is to get the right answer as a result of having only half of the data in hand, and half of it wrong, and you don't know which half was wrong. When you get the right answer under those circumstances, you are doing something creative, which is certainly what happened in the instance described.

Photochemistry

My work on the photochemistry of the triplet state, together with my long-standing interest in porphyrins, led to the discovery that the photooxidation of chlorin was the result of the transition of the excited chlorin into a triplet with a long enough life that it could be oxidized.[38] Our study on the photochemistry of porphyrins[39] was well on its way.

In connection with our study of the effect of structure on the mechanisms of reactions, which began with Gerry Branch, we had looked at the effect of various substituents on the restriction to rotation around the double bond, particularly in stilbene.[40] As mentioned elsewhere, when we made the 4-dimethylamino-4'-nitrostilbene it was impossible to get isomers around that double bond, because the resonance structures so induced reduced the restriction to rotation around the double bond so much that at room temperature only one compound was isolable. At first we were measuring the formation of the *cis*-stilbenes by illumination of the *trans*-stilbenes, and then measuring the thermal reaction in the reverse direction of the *cis*- back to the *trans*-stilbene. When we tried to do this with *o*-nitro-substituted stilbene (that is, stilbenes in which there was a nitro group in a position ortho to the double bond), we found an entirely different sequence of events. The photochemistry of such a compound was very complex, leading to a complete rearrangement of the skeleton, first into an isotagen and then into a nitrone (Scheme III). This work was described first in 1955 and is continuing today.[41]

During our efforts to elucidate the path of carbon, we also studied the photochemistry of a cyclic disulfide, thioctic acid (a five-atom ring disulfide). This compound had just been discovered as a new cofactor in biological oxidation–reductions, and it seemed to me that it might conceivably be playing a role in the initial separation of oxidizing and reducing power as a result of the light absorbed by the chlorophyll.

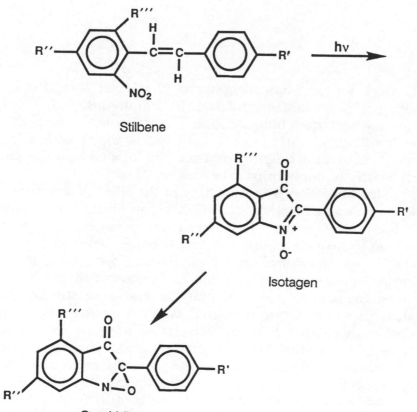

Scheme III. Rearrangement of a stilbene into an isotagen and then into a nitrone.

I obtained a sample of thioctic acid from T. H. Jukes, then at Lederle Laboratories, and found that its absorption spectrum was indeed very different from that of ordinary disulfides. It had an absorption stretching into the visible range, whereas the absorption of glutathione and other larger-ring disulfides was found in the ultraviolet range. I attributed this feature to the fact that in a five-membered ring disulfide the dihedral angle must be close to zero and thus under strain, raising the level of the ground state and moving the absorption toward the visible range. I also thought that the excited state of the thioctic acid would react with water in a hydrolytic cleavage, giving an SH group on one sulfur atom and a SOH (sulfenic group) on the other. Thus, one of the sulfur atoms would become a reducing agent and the other an oxidant, achieving the principal result of the primary quantum conversion in photosynthesis.

We spent a great deal of time and effort examining this reaction experimentally,[42] and indeed the thioctic acid performed much in this way. However, the most sensitive biosassays that we could perform at that time did not demonstrate its presence in photosynthetic tissue. Thus, thioctic acid turned out to be a "blind alley" as far as the photosynthesis work was concerned. After several years we gave up our study of the photochemistry of disulfides.

Electronic Properties of Organic Crystals

One of the studies related to the quantum conversion process that we undertook in the late 1950s was the elucidation of the electronic properties of organic solids, particularly the porphyrins. We examined other materials as well, such as the stacked condensed ring systems of polycyclic aromatic crystals.

However, our principal interest was in the solid porphyrins, with phthalocyanine as an example. This study of the organic solid state took two forms. One of them involved the study of the induction of conductivity by illumination of the solid crystals[43] and the other involved the effect of added impurities on the conductivity and the photoconductivity of these solids.[44] Part of the reason for examining such solids, which, in effect, were ordered arrays of organic molecules, stemmed from our conviction that it was such ordered arrays that gave rise to the very special properties of photosynthetic systems.

We found, of course, the existence of conductivity in these stacked aromatic crystals, which was enormously enhanced by adding suitable impurities, usually electron acceptors such as a trace amount of quinone in a polycyclic aromatic compound or phthalocyanine. Even the presence of oxygen on the surface of such crystals gave enough conductivity to the mass by removing electrons from the mass, thus leaving behind holes that could conduct charge. However, this type of a system is not a good model for what occurs in the organized photosynthetic apparatus of the green plant.

Chemical Evolution and Organic Geochemistry

In 1951 our experimental work on chemical evolution resulted in a short note[45] in *Science*. This work stemmed from two sources. One was the course in paleontology that I took as an undergraduate student, and the second was the occasion in 1949 when I had spent some time in the hospital. During that period I had read the book by George Gaylord Simpson entitled *The Meaning of Evolution*.[46] This combination of circumstances, together with my concern for how the complex system of photosynthesis had evolved, led me to the question of how the initial organic materials might have formed on the surface of the earth. Photosynthesis itself was much too complex to have been the initial route to organic materials.

The experiment that resulted in the 1951 publication[45] was performed on the 60-inch cyclotron at Crocker Laboratory. An aqueous solution of CO_2 and hydrogen was irradiated with accelerated helium ions to give rise to simple reduced carbon compounds such as formic acid, a bit of formaldehyde, and some oxalic acid, for example. This work led to a discussion with Harold Urey in 1951 or 1952 that took place in the old Radiation Laboratory. He argued that the early atmosphere must have been reducing, rather than the oxidized one (CO_2) that we had used. In our initial experiment we did not have any nitrogen in the simulated prebiotic atmosphere. Several years later Stanley Miller, then at the University of Chicago, and Professor Urey performed a similar experiment in which the gaseous system contained methane and ammonia with water. This combination led to the formation of amino acids.[47]

Our experiment in 1951 was the beginning of many studies done in many laboratories throughout the world (incidentally, they are still continuing) for the purpose of determining the first organic compounds created on the surface of the earth by various types of energy inputs.[48] The efficiency of the conversion of the simple to the more complex compounds clearly depends on the redox state of the initial gas mixture and is higher from the more reduced initial gas. The reason for this dependence is that when we start with CO_2 in the reaction, its reduction must occur prior to any further formation of carbon–carbon bonds or even carbon–hydrogen bonds. My guess is that had we done our experiment with added nitrogen, we would have seen nitrogen-containing compounds as well, even starting with CO_2.

As time went on, a larger number of the monomers that are important for biopolymers appeared in the irradiated mixtures that simulated the atmosphere of the prebiotic earth. For example, adenine appeared quite early,[49] as well as some of the higher amino acids in addition to glycine and alanine.[50] Finally, with the appearance of HCN and then cyanamide[51] and dicyanamide,[52] the means of a dehydration polymerization of these monomers was at hand.[53] Soon polypeptides and polynucleotides appeared in these irradiated mixtures. Thus the molecules necessary for the formation not only of the cell membrane, in the form of the surfactant fatty acids, but also of its apparatus[54] in the form of the biopolymers,[55] peptides, polynucleotides, and polysaccharides were present.

In this mixture the porphyrins had also been formed, and the trace elements such as iron were always present, so the autocatalytic systems for oxidation and then, finally, of synthesis began to evolve. As soon as stereospecific autocatalysis appeared,[48] the very characteristic reactions of living organisms could be selected for. One such asymmetric catalyst type that could perform such asymmetric reactions would be the properly substituted cobalt-containing amines.[56]

In fact, an asymmetric autocatalysis has been experimentally observed in the case of allyltrialkylammonium ion. When the three alkyl groups, each different from the other and different from the allyl group, are present, the dissolved quaternary compound is in rapid equilibrium between the two optical antipodes.[57] However, as the solution evaporates, crystals of the quaternary salt begin to appear. Apparently the crystals are those of either the D- or L-form, and after one crystal is formed of either of those enantiomers (randomly) it nucleates (catalyzes) all the subsequent crystals to come out in the same form. Thus, most of the quaternary salt will be of one optical form spontaneously resolved. The choice of optical isomerism in such an experiment, done independently several different times, seems to be random, but once the selection is made in any one experiment all the subsequent material appears in that particular form.

Once such stereospecific autocatalytic systems evolved, the whole complex of organic molecules that had originally been randomly formed would rapidly disappear in favor of the more specific ones that were now possible and selectable.[58]

Our chemical evolution studies were extended to the examination of ancient (Precambrian) rocks for the presence of "molecular fossils" (chemical fossils) or, if you like, the residues of molecular structures that might be found in ancient rocks and would provide evidence of living organisms at that geological period.[59] This same approach was extended to the examination of meteorites from outer space to see if there was any molecular evidence for the extraterrestrial existence of life.[60] This turned out not to be so (no evidence). Our first paper on the subject of space chemistry was published in 1960, and in 1964 our first paper in organic geochemistry appeared;[61] that work continued until 1978.

In our search for evidence of ancient life we soon discovered, in some of the oldest rocks we could obtain for examination, the presence of phytane and pristane, as well as a number of other isoprenoids as hydrocarbons. These seemed to us to be unambiguous evidence of the existence of organized living things that could produce such highly organized organic compounds.[62] An ordinary irradiation or thermal cracking of materials like methane, together with other simple compounds

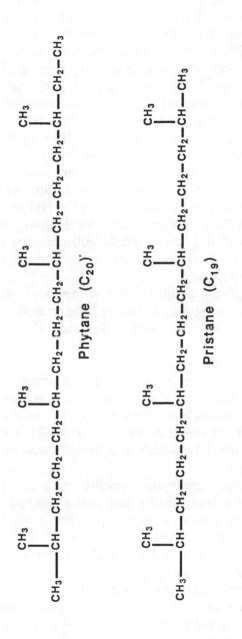

Phytane (C_{20})

Pristane (C_{19})

such as ammonia, might give rise to a random collection of structures, but not these ordered and periodic arrays represented by the isoprenoids, phytane, and pristane.

In addition, we began to examine ancient paleontologically defined plant materials and were able to discern the same kind of evidence.[63] In order to confirm that these residues might be those of ancient algal mats, we examined modern algal mats, ranging from a few hundreds to thousands of years of age. These are the recognizable residues of dead algae that have been matted together by virtue of the large amount of cellulose in their structure, one seasonal layer upon another. The material has only just begun to be transformed into what will become kerogen in the ancient fossilized mats.[64] The modern mats were obtained from Baja California and local blue-green systems. In them we indeed did see a whole collection of even-numbered fatty acids, as well as isoprenoids of the general character found in the ancient rocks.

Although we sought other components that might have survived such a long and tortuous history as these ancient rocks went through, it became obvious that such classes of compounds as amino acids, peptides, nucleic acids, sugars, and other more complex structures could not have survived, and that the only ones that had a reasonable chance of surviving a billion years or so of modification that the rocks went through would be the hydrocarbons themselves. Random polymerization of hydrocarbons such as methane also was found, which would result in a broad continuum when such a collection was passed through a gas chromatograph.

This broad continuum results from the fact that when such a reaction is undertaken, one gets almost all possible combinations of CH_2 groups—chains and branches, including most isomers. At one point in the history of the rocks we could find not only a continuum of hydrocarbons but also, superimposed on that continuum, sharp peaks due to specific straight-chain unbranched hydrocarbons. Such components must have had their origin in a straight-chain fatty acid by simple decarboxylation of the biologically formed even ones, especially because the straight chains that were found superimposed on this continuum were, in general, odd-numbered hydrocarbons.[65]

These organic geochemical studies eventually led to examination of the organic constituents of moon rocks, which we received in 1969, to show the molecular evidence for "life" in outer space.[66] So far as our work was concerned, there was no evidence of that kind, and I believe no one else has found it either.

The studies in chemical evolution and organic geochemistry were discussed in a book published in 1969 entitled *Chemical Evolution: Molecular Evolution Towards the Origins of Living Systems on Earth and Elsewhere.*[67] This book resulted from a series of lectures I gave at Oxford University, England, while I was George Eastman Visiting Professor in Balliol College (1967–1968).

Free Radicals

We had electron spin resonance (ESR) equipment available here in Berkeley (one of the earliest of these machines), and were able to demonstrate the generation of unpaired electron spins by illumination of photosynthetic systems as more attention was paid to the photochemical part of photosynthesis. The first observation of unpaired spins induced by light in photosynthetic systems was in 1957.[68] The ESR equipment with the spin resonance detector device was used to examine many other organic reactions, and we were able to show that a number of organic reactions proceeded via free radicals.[69]

With the availability of the electron spin resonance detector, we did another experiment in which the hyperfine structure of a semiquinone free radical was observable. This semiquinone free radical was the one made from tetrafluoroquinone, and it showed the five-line hyperfine spectrum induced by nuclear splitting from the four fluorine atoms. This spectrum was very easily seen when the semiquinone concentration was on the order of 10^{-4} M. As excess quinone was added to this semiquinone solution, the hyperfine structure was gradually broadened. At the concentration of about 10^{-1} M quinone, the hyperfine structure was completely obliterated, and all that was left was a single broad resonance line.[70]

This result was interpreted in terms of the rapid exchange of the unpaired electron between the semiquinone molecules and the excess quinone molecules that were present such that the orientations of the fluorine atoms were averaged out. This rapid exchange in 10^{-1} M solutions has become an important idea in planning to develop a system for electron transfer through an otherwise insulating medium. This original work on the hyperfine structure was done in 1962, but the work following from it toward the development of a photochemical electron-transfer system through a solid polymer loaded with a suitable number of electron-transfer sites is only now being devised.

In 1961 one of our first papers on primary quantum conversion in photosynthesis appeared,[71] a study in which we are still engaged. Work in this area has continued in many different directions, including free radical production during the primary quantum conversion act, determination of the various light reactions involved, and our current emphasis, photochemical conversion and storage of solar quanta (*see* the later section on artificial photosynthesis). This basic project might best be categorized under the general heading of "photochemistry", but it includes many subdivisions and pathways that are not strictly photochemical in nature. In this connection, the term "quantasome" was devised for the small particles in green plants that constitute the photochemical center.

Effect of Deuterium
on Biological Reactions

As a consequence of our earlier observations on the isotope effect of deuterium in chemical reactions, particularly concerning the strength of hydrogen bonds affected by deuterium, we thought of the possibility of using this deuterium-induced change in hydrogen bond strength to effect the replication of viruses, which by that time were known to be essentially nucleic acid structures held together by hydrogen bonds. It seemed to me that if we could replace the protons in these structures with deuterium we might inhibit the growth of viruses and thus have a system for treating viral diseases.

In order to test this theory, we infected mice with a virus and then put them on a D_2O diet. It turned out that we could get the deuterium content of the mouse's body water up to about 18–20%, which was enough to inhibit virus replication; beyond that concentration, the mouse itself would be destroyed.[72] In performing these mouse virus experiments with D_2O, we did not separate the sexes of the mice. We kept the mice in the same cage, both the males and females. It was noticed that the deuterium-treated mice, while they were cured of the virus, were sterile; that is, the mice could produce no young.[73] The control group, in which the mice did not receive deuterium, produced a litter periodically. Detailed examination of this phenomenon showed that the male mice were sterile when they were deuterated, presumably due to the failure of mobility of their sperm.

In the 1930s Gilbert Lewis did some similar experiments, feeding some D_2O to a mouse.[74] The mouse did, however, survive, but I believe the reason was that Lewis ran out of D_2O (he had only a few milliliters); the mouse would not have survived if Lewis had increased the dosage over a long period of time.

When it became clear that heavy isotopes of hydrogen existed, Lewis realized that deuterium would be the easiest one to separate. He managed to obtain material containing about 70% deuterium by using water from the electrolytic cells for the production of hydrogen gas that Professor William F. Giauque had been using in Gilman Hall. Eventually Lewis obtained water with over 95% D_2 and he used this material for both physical and biological experiments.

At about the time we were doing the mouse experiment (1958), I went to Cornell University to give the Baker Lectures. At Cornell there was a detailed study of helical proteins underway by means of melting point studies. I felt it would be worthwhile to examine the melting point of a helical protein, which is presumably held together by hydrogen bonds. When the protein is deuterated it should change its melting point; this turned out to be a correct assumption. The melting temperature of the deuterated protein was reduced by about 10 °C.[75]

Chemical and Viral Carcinogenesis

In the early 1970s we returned to the work on chemical carcinogenesis that had been started in 1946 by the synthesis of ^{14}C-labeled dibenzanthracene by Charles Heidelberger.[76] Various disciplines were involved, ranging from synthetic organic chemistry to cell biology.

We synthesized compounds of particular interest in cancer research, determined the molecular characteristics of carcinogenic hydrocarbons, performed mechanistic studies on chemical carcinogens, tested them in tissue culture, and modified their organic structure when it seemed feasible to do this.[77] The work led to the identification of the active metabolite of at least one polycyclic aromatic hydrocarbon, benzo[a]pyrene, as its diol epoxide.[78]

With the availability of the specialized organic compounds, the work immediately expanded into viral carcinogenesis as well.[79] At that time we realized that chemical carcinogenesis and viral carcinogenesis are synergistic. When we measured the rate of chemical transformation on a tissue culture and a similar rate of viral transformation on a corresponding culture, we noted that if the chemical treatment is introduced first, followed within the repair period by the addition of virus, the carcinogenesis is enormously enhanced—not just additive, but enormously enhanced.

This enhancement suggested to me that the chemical was damaging the DNA of the cell. During the course of cellular

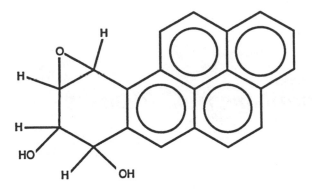

Benzo[a]pyrene Diol Epoxide

DNA repair, any viral oncogenes that were present had a chance of being integrated, thus giving a much larger transformation rate. Our first paper on the concept of synergism between viral and chemical carcinogenesis was published in 1973. Since that time the entire idea of oncogenes stemming either from virus particles, plasmids, or even in the genome itself by rearrangement has been fully established.[80] It is now the principal theory of how cells are transformed into malignancy.

Following the notion that all carcinogenesis involves the integration of an endogenous piece of DNA, particularly viral DNA, the idea arose of using a drug with antiviral properties as an anticancer agent. Such a drug was available to us in the form of rifampicin, which was known to have antiviral properties, especially as an inhibitor of RNA-dependent DNA polymerase and so should be active on onco RNA virus. A trial with this drug on a fully established tumor was not very successful. However, these experiments did show pronounced inhibition of the transformation process itself, by both chemical and especially viral carcinogens.[81]

Hydrocarbons from Plants

As a result of the political and economic effect of the 1973 oil embargo, we began our effort to find plants that would produce hydrocarbons directly, and our first paper on this subject was published in 1974. This gave rise to our whole activity on the generation of important new fuel sources by using plants to convert solar energy into hydrocarbons on an annually renewable basis, an area of research that constitutes one of our two major activities today.[82]

Some plants store the sun's energy directly as hydrocarbons. The most important commercial plant of this type is the rubber tree (*Hevea*), which has a low mass efficiency although it produces hydrocarbons directly. These hydrocarbons are not practically useful for fuel production because of their high molecular weight. Material from *Hevea* has other more useful properties, namely, the elastomeric properties that generate rubber. Rubber is a commercial hydrocarbon crop, harvested annually. The yield of that crop has been improved by a factor of 10 over a short time.

By changing agronomic practices and improving plant breeding, the rubber growers raised production to 2000 pounds per acre per year, which is the commercial productivity of the Malaysian rubber plantations. There are even small plots of *Hevea* that produce 4000 pounds per acre per year, and individual trees, when extrapolated to plantation size, that produce 8000 pounds per year, which is equivalent to 4 tons of hydrocarbon. Because 1 ton of hydrocarbon is the equivalent of about 7 barrels of oil, this means that the equivalent of 25 barrels of oil per acre per year can be obtained from hydrocarbon-producing trees. The Malaysian rubber producers began with an equivalent of only about 3 barrels of oil per acre per year, and they could eventually produce the equivalent of about 30 barrels of oil per acre per year. This is an example of what can be done with biological material when you begin with wild plants and breed and cultivate the plants under the best possible conditions.

Posing with a Euphorbia lathyris *plant in the garden of my Berkeley home.*

It seemed to me that a systematic effort to examine the latex from various plants and analyze it for the hydrocarbon content might be one way to establish whether these plants could be used on any large scale as a source of renewable materials. Therefore, we (my wife and I) began our search for plant latex from species other than *Hevea*, although initially we used the latex from the rubber tree as a standard material in our analytical work to devise the methods of extraction, analysis, chromatography, etc.

Collaboration with Genevieve Calvin in Energy Plantation Research. When the idea was developed that plants could be grown for their hydrocarbonlike materials, it became important to locate the sources of plants of that type not only in California but elsewhere throughout the world. It was no accident that my wife, Genevieve, called my attention to the existence of a particular plant of this type, *Euphorbia lathyris*, commonly called gopher plant because of its ability to control gophers in the garden. She had cultivated these plants in our Berkeley garden. An examination of the botanical literature revealed that these Euphorbias, like other species of *Euphorbiaceae*, extruded a latex that could be extracted for hydrocarbonlike chemical compounds.

On our ranch in Healdsburg, about 85 miles from Berkeley, Genevieve herself began the cultivation of a "petroleum plantation". At one time she had about 5 acres under cultivation. These plants were used as a source for seed, which was sent to researchers throughout the world who wrote to the laboratory requesting seed for further experimentation.

We traveled throughout the world seeking plant species that would be useful for energy and materials. We went to Brazil, Malaysia, Spain, Canary Islands, many European countries, and Puerto Rico, as well has having discussions with farmers in the Midwest and Southwest of the United States. There is no question that Genevieve's enthusiasm for this effort was a great impetus for me to continue the experiments, discussions with the Department of Energy for funding, and our worldwide contacts to discuss the possibility of using plants for energy and materials.

Genevieve and I at our ranch in Healdsburg, California. The Euphorbia lathyris *plantation is in the background. (Reproduced with permission from reference 82f. Copyright 1981.)*

During the course of a trip to Brazil, which had been organized by the Dow Chemical Company, I met a sugar cane planter by the name of Evaldo Inojosa. His plantation was situated east of Rio de Janeiro, near a town called Campos. I do not recall the precise occasion of our first meeting, but the product was a warm friendship between our families that resulted in repeated trips to Brazil under the sponsorship of his sugar company when I addressed the annual meetings of the sugar cooperative, COPERFLU, of the State of Rio de Janeiro.

These trips usually involved a stay of 2 weeks or more at the Inojosa plantation. The Inojosa home was a large one-story building surrounded by an outdoor screened porch on which most of the social activities of the family took place. When we first met Evaldo and his wife, Maria Angela, they picked us up at the Rio airport, and we could barely communicate with each other because we knew no Portuguese and they knew very little English. However, on the trip from the airport to the plantation we did manage to explain to each other who we were, and we got to know the entire family, including the children and various brothers, sisters, and in-laws.

During our stays with the Inojosas, which were repeated over a period of many years, we also became acquainted with other sugar planters, all members of COPERFLU, who lived in the vicinity of Campos. Each had a large plantation and a large house. The house and establishment around it seemed to me to be very much like my impression of the style of a pre-Civil War cotton plantation in the South of the United States.

Part of the interest of the sugar planters was based on the possibility of their producing another product—namely, fermentation alcohol for fuel—from their sugar. They had produced small amounts of it from the supernatant of the sugar recrystallization. However, as a result of my discussions with them, both formally and informally, the idea of fulfilling part of the liquid fuel needs of Brazil through large-scale alcohol production from sugar cane was born. Evaldo took me to visit the minister of energy in Brasilia and eventually the president of Brazil, and we talked about the possibility of large-scale alcohol production for fuel. I believe this eventually resulted in the Brazilian ProAlcool program which has led to the construction of many "autonomous" alcohol-producing plants. These were fermentation plants constructed for the direct fermentation of the total sugar from the cane to alcohol and not for sugar products. This development gave rise to the large amount of fuel alcohol that was ultimately used in Brazil.

Fuel and Materials from Plants. The hydrocarbon in most latex-producing plants represents about one-third of the total latex. Therefore, if we could find a family of plants that contained latex, it might be a fruitful avenue for chemical research. In the genus *Euphorbia*, a member of the *Euphorbiaeceae* family, almost every species is a latex-producing plant. This fact could be interpreted to mean that the energy of the sunlight is stored in these plants as hydrocarbon.

If the plants produce hydrocarbonlike materials and if the yields can be improved materially, perhaps other latex-producing plants can be found that can be grown under less humid and tropical conditions, perhaps on land that would actually be arid or semiarid, such as the southwest of the United States, the African deserts, the desert areas of Chile, Argentina, and Brazil, and many areas in India. In addition to making the land itself productive by suitable selection of plant species, it might be possible to produce on that land material such as hydrocarbons that would have direct economic use.

This idea led to a trip to Brazil in 1975, the first of many (*see* preceding section), where we found hundreds of species of *Euphorbia*, which were latex producers, and some of them grew in relatively arid regions. It turns out that some of the species of *Euphorbia* that grow in the arid regions of Brazil, such as the San Francisco River basin, can also grow in the arid regions of the United States. One of these species, *Euphorbia tirucalli* (milk bush) yields a latex that can be harvested by tapping, although

Euphorbia tirucalli *in Brazil.*

it can also be harvested by cutting the same way that sugar cane is cut. To produce "oil" from the *E. tirucalli* the fleshy stems could be crushed, the latex extracted by suitable chemical processing, with the recovery of the solvent. Because of the characteristics of latex production in the leaves and its ability to grow in dry areas, the *E. tirucalli* appeared to be an excellent candidate for a "gasoline tree plantation".

About 40 years ago a species of *Euphorbia*, *E. resinifera*, was grown in Morocco. Approximately 125,000 hectares of land were harvested with a production of 10,000 liters of latex per hectare, which resulted in 1700 kg of rubber (benzene extractables) and 2750 kg of gum resin (acetone extractables). It is not known whether this effort was ever repeated or whether it was done once only. It does, however, reinforce the feasibility of growing *Euphorbia* for the latex it produces. Also, the Italians endeavored to create *Euphorbia* plantations in Ethiopia in the 1930s, but this effort was abandoned because of the war there in 1936.[82]

Another genus that might be useful for renewable resource production is the *Asclepias* (milkweed), which will grow in most areas of the United States. In Brazil the milkweeds grow to a height of 8–10 feet. The parts of this plant also could be crushed and extracted for their hydrocarbon content, in a manner similar to the parts of the *E. lathyris* and *E. tirucalli*. In fact, plantations of *Asclepias speciosa* have been developed in Utah and work has been done on trying to improve the yield of hydrocarbonlike materials from the harvested plants.[83] In 1977 on a trip to Puerto Rico we also discovered that *Asclepias* grow there profusely in the dry regions of the South Coast. That species would obviously be an excellent candidate for hydrocarbon production via plants.

In our laboratory we examined about a dozen species of *Euphorbia*, and most of them contain hydrocarbons of a much lower molecular weight than rubber.[84] The acetone extracts of the leaves of the hydrocarbon-containing plants contain mostly sterols, which can be separated by gas–liquid chromatography coupled to mass spectroscopy to identify the individual sterols (Figure 2). Sterols of five different plants are shown, as are the numbers of the various peaks that identify the sterols (C_{30} compounds) with molecular weights on the order of 500.

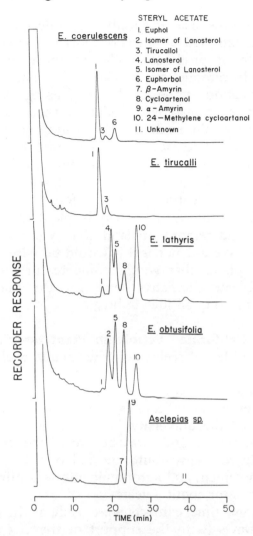

Figure 2. GLC of Euphorbiaceae *steroids. (Reproduced with permission from reference 86b. Copyright 1988.)*

During the course of the next several years, plantations of *Euphorbia* were developed in California and elsewhere in the world as well (specifically in Spain, Japan, and Thailand). This effort entailed the development of extraction procedures and analytical techniques for the various plant components. We learned through the use of specific catalytic materials such as zeolite that the extracted products of *Euphorbia lathyris* cracked to the usual suite of products similar to those obtained when petroleum hydrocarbons are cracked.[83,86] This finding was another confirmation of the feasibility of using plant hydrocarbons as a source of materials similar to those produced in a petroleum refinery. We made an effort throughout the world to find plant sources of hydrocarbon that could be extracted and converted into chemicals and fuels to be used in place of the residues of ancient photosynthesis (coal, oil, and natural gas).[87]

The essential feature of our effort to find renewable energy resources is that we have sought plants that would grow in arid or semiarid regions on land that is presently relatively nonproductive. We did not feel it would be wise, or even politic, to explore plants that would compete for food-producing land, although an alternative crop for the grain-producing Midwest might be a good idea today.

University of California "Petroleum Plantations". The initial impetus for the idea of seeing whether plants could be used as sources of hydrocarbonlike materials came as a result of the 1973 gasoline embargo, as previously mentioned. After discussing this situation with my colleagues, I went to the vice-president for agriculture of the university for "seed" money to get some agronomic experiments underway at two of the university field stations, one in southern California and one in San Jose. The proposal was accepted, and agronomic efforts continued at these two field stations for about 3 years.

At the same time, contacts were made at the University of California at Davis (with the support of the vice-president for agriculture) with people knowledgeable in agronomics. At Davis several acres of *Euphorbia lathyris*, the first plant we studied, were planted, and various parameters (salinity, water quantity, pesticides, etc.) were measured. The final conclusion of this work at Davis about 8 years ago was that this type of "plantation" was not economically viable, partly because the actual yield of hydrocarbonlike materials was not high enough to justify the development costs for the extraction machinery, marketing efforts, and other factors (such as the then-low price of oil).

It is my personal view that the effort to develop crops of hydrocarbon-producing plants should be continued by having farmers in various parts of the country use their set-aside land (from government subsidies) to grow experimental plants for fuel and materials content. However, until the federal government provides some encouragement for this type of an effort, it is doubtful whether anything significant will develop. However, a great deal of basic information about the biosynthetic routes of hydrocarbon-producing plants has been obtained and may be used at some future time when the political and economic climate will be more amenable to suggestions of this type.

Biosynthetic Pathways in Hydrocarbon-Producing Plants. Today in our laboratory the emphasis has changed to determining the organic and biosynthetic pathways of terpenoid production in plants, with a view toward possibly increasing the efficiency of various steps in this sequence to produce a greater quantity of usable plant products.[88] For example, it appears that the hydroxymethylglutaryl coenzyme A reductase may be the limiting enzyme in isoprene synthesis. Various species of plants, including trees, are being explored to find if there is some combination of genes for isoprenoid production in the biosynthetic pathway(s) that could combine the terpenoid production of one plant species with the desirable growth characteristics of another species. If we knew the rate-limiting steps, we might have a chance of introducing suitable genes to increase the rate of hydrocarbon synthesis.

Artificial Photosynthesis

We also began to think, in the early 1970s, in terms of totally artificial systems that could capture solar energy and convert the quanta into useful products. We began to call our first artificial photosynthetic systems "synthetic chloroplasts" because their structure simulated the plant chloroplasts themselves.[89]

We realized that we would need very special systems in order to accomplish quantum conversion in a totally artificial system with anything like the efficiency of the natural system in green plants. We examined the structure of the materials in the green plants that actually do the quantum conversion and came to the conclusion that the factor that made it possible for the green plant to be so successful was that, in every case, it appeared that the photoelectron transfer occurred across a phase boundary.[90] In 1971 we did our first experiments on sensitized electron transfer on semiconductors, in which chlorophyll was used as a sensitizer on zinc oxide.[91] It was possible to demonstrate that excited chlorophyll transferred an electron to the zinc oxide and the electron current flowed in that direction; the chlorophyll could then be regenerated by a suitable reducing agent in the medium in which the sensitized semiconductor was immersed.

We then set about constructing artificial photosynthetic systems by simulating this quantum conversion process on phase boundaries more generally.[91] The first phase boundary we examined was a semiconductor–water surface, and this was successful to a degree. However, it seemed that it would be better to produce our sensitizers, donors, and acceptors on membrane to separate the oxidized product from the reduced product so the back reaction would be inhibited. This notion led to the construction of surfactant sensitizers analogous to chlorophyll and of suitable donors and acceptors that could be placed on either side of the membrane in order to achieve a satisfactory electron transfer across that membrane (Figure 3).[89]

The synthesis of surfactant porphyrins and other surfactant sensitizers,[92] such as a derivative of trisbipyridylruthenium, then took place, and we began to incorporate these and other similar materials into our membranes. The membranes we used were usually constructed by using either a natural phospholipid or a synthetic surfactant lipid that would spontaneously form

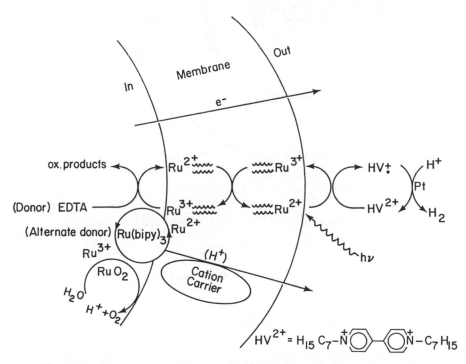

Figure 3. Electron transfer across lipid vesicle wall. (Reproduced with permission from reference 89. Copyright 1983.)

bilipid membranes in the form of spherical bilayer vesicles.[93] In these spherical systems a solution was thus contained in the center of the sphere and was separated from the water in which the sphere was suspended by the spherical membrane itself (Figure 4).

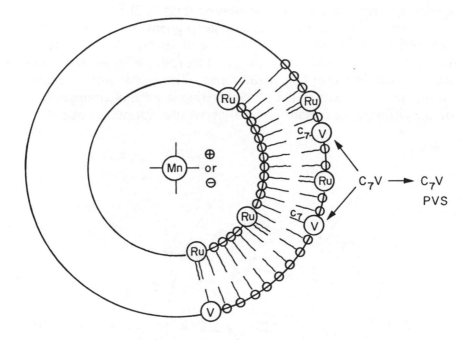

Bis Hexadecyl PO_4 (Phospholipid)

$: Ru^{+2}(bipy)_2\big[bipy(C_{16})_2\big]$ (~100:1)

10^{-2}–10^{-3} C_7V + PVS outside

10^{-4} $MnP^{(+ \text{ or } -)}$ inside

pH 9–10 both sides

Figure 4. Single unilamellar vesicle. (Reproduced with permission from reference 89. Copyright 1983.)

This effort to construct synthetic chloroplasts, or artificial photosynthetic systems, has led to a major effort in the photochemistry of surfactant sensitizers. A variety of systems has been constructed with this basic idea in mind. The two systems that we have used the most have been the synthetic tetrapyridylporphyrin in which one of the pyridyl groups has been alkylated with a long chain, thus making it into a surfactant porphyrin. The other has been the trisbipyridylruthenium in which we have attached two long-chain hydrocarbons to one of the bipyridyl groups so that it, too, will be a surfactant trisbipyridylruthenium.

Each of these normally takes its place at the water—oil interface of a vesicle or micelle. Later a much simpler phase boundary was used. The sensitizer and electron donor were adsorbed on inert negatively charged silica particles, and an acceptor was constructed that was neutral but became negatively charged following the electron transfer. The presence of the highly charged silica colloid prevented the back reaction, and relatively efficient storage could be obtained.[94]

Successful electron transfer across a membrane has occurred with the sensitizers described. In general, we have used both irreversible acceptors for one side of the membrane and/or irreversible donors for the other. As yet we have not placed catalyst systems on both sides of the membrane to generate hydrogen, or reduced carbon dioxide, on one side and oxygen on the other. We hope this will be done when we have found the proper catalysts and electron relays.

Because we knew that manganese was involved in the oxygen evolution step in photosynthesis, we began in the early 1960s a study of the chemistry of manganese complexes, in the first instance porphyrins because they were easy to prepare and stable.[95] We have succeeded in oxidizing the manganese porphyrin from the Mn(III) state in which it normally exists into the Mn(IV) state that is not stable and reverts to Mn(III) in pure water.[96] It is not yet clear that this reduction reaction of manganese can make oxygen, but something like this must be happening (for example, an oxidation of the porphyrin itself). Obviously a more stable structure to hold the manganese must be found, perhaps phthalocyanine. On the other side of the reac-

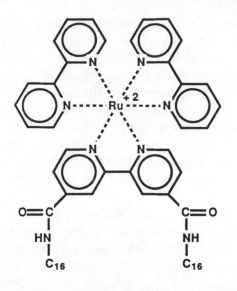

surfactant   Ru(bipy)$_2$[bipy(CONHC$_{16}$)$_2$]

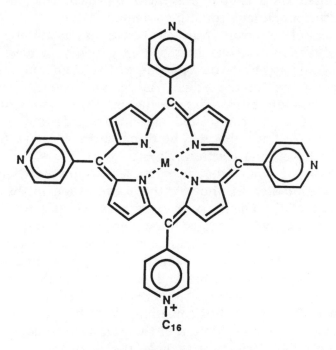

surfactant   porphyrin

tion, as an irreversible acceptor, we have used chloropentam-minecobaltic chloride, which, upon reduction to the cobaltous state, immediately decomposes into cobaltous ion and ammonia; thus, no reverse reaction is possible.

This system has also been successful in demonstrating photosensitized electron transfer. We have used several violo-gens as electron acceptors, which then, with a suitable catalyst such as platinum, will produce molecular hydrogen. We can also use ruthenium oxide, a heterogeneous catalyst, on the oxi-dized side of the system so that the oxidized manganese or the oxidized ruthenium, which would be formed on that side, would also generate oxygen. However, the use of heterogene-ous catalysts in these microvesicle systems has some difficulties, particularly the catalyst particle size, and we continue the search for homogeneous catalysts for the hydrogen and for the oxygen production.

We are attempting to learn the principles of photochemi-cal conversion and storage of solar energy so that it will be pos-sible to construct synthetic devices to perform the photolysis of water, thus producing molecular hydrogen or reduced carbon dioxide[97] and useful oxidation products.[89]

Celebrating my birthday, April 8, 1989. Left to right, back row: Larry Spreer, Marilyn Taylor, John Otvos, Scott Taylor, and Janet Splitter. Front row: Rich Lumpkin, Carl Craig, the guest of honor, and Ben Gordon. This group is currently involved in research on arti-ficial photosynthesis.

Organization of the Laboratory of Chemical Biodynamics

As mentioned earlier, the Bio-Organic Chemistry Group of the Radiation Laboratory at Berkeley had its beginning in December 1945. This group was the nucleus of what later was to become the Laboratory of Chemical Biodynamics of the University of California and the Chemical Biodynamics Division of the Lawrence Berkeley Laboratory.

The original scientific staff of the Bio-Organic Group consisted of six senior Ph.D. chemists, each with a specific expertise. These men acted in concert to keep the laboratory group functioning. Decisions about the direction of the research, personnel, and funding administration were made at a weekly luncheon meeting; the basic path of the laboratory activity was set at these gatherings. In spite of the varying backgrounds of the participants, a harmonious consensus of what would be best for the entire laboratory, rather than for individual research programs, kept the organization on the cutting edge of science for more than 35 years.

After some years of operation, several other faculty members from the Department of Chemistry joined the initial group of scientists, and that trend continued. At the time I retired as director in 1980, there were five chemistry faculty members, one faculty member from botany, and eight senior scientific staff members who directed the research effort of the postdoctoral associates and graduate students.

This method of operation of a large, complex research group with many different disciplines led to a continuity of research style that existed for more than 25 years. The people who came to the laboratory as postdoctoral associates and graduate students eventually went on to situations of their own, where they established research groups that reflected the interdisciplinary background they had observed in Berkeley.

At the beginning, we were physically separated into two groups, one group, the carbon-14 synthetic and the animal biochemistry group, being housed in the Donner Laboratory, and the photosynthesis research group, housed in the old Radiation Laboratory (ORL), an old wooden structure from which

In our "deerstalker" hats on the steps of the old Radiation Laboratory, 1952, are, from left to right: J. Rodney Quayle, Vice Chancellor, Bath University, England; Richard E. Norris; R. Clinton Fuller, Department of Biochemistry, University of Massachusetts, Amherst; Malcolm Thain, Tropical Products Institute, London, England; and me. I don't remember the impetus for getting the hats, except that there were two Englishmen in the laboratory and we were "stalking" the elusive two-carbon fragment in the photosynthetic carbon cycle. The deerstalker caps were ordered from London, and I, at least, still have mine and wear it in the wintertime.

the original 37-inch cyclotron developed by Ernest Lawrence had just been removed. The room where the 37-inch cyclotron had been situated was a large open space, a laboratory without internal walls. We installed laboratory benches, each capable of holding two people on each side, and partitioned off a small corner that we used as office space. It was necessary to pass through the office to go outdoors and downstairs to the basement of ORL underneath the wooden building, in which we had built a concrete "house" for the Geiger counters to reduce the background radiation to a minimum. Also constructed underneath the building at a later date was another concrete room that housed our earliest nuclear magnetic resonance and electron spin resonance equipment.

Architect's rendering of the Laboratory of Chemical Biodynamics, University of California.

This type of laboratory space—that is, an open laboratory with no walls separating the benches—gave rise to a degree of interaction between the scientists of diverse disciplines who were working there, which was special in very many ways. Everyone knew pretty much what everyone else was doing and could participate in discussion freely, as well as help in an experiment if need be.

Therefore, when the time came for us to design a new building for our work, we did it so the structure itself would have built into it the principles of interaction that we had discovered quite by accident in the old days in ORL. The design of the Laboratory of Chemical Biodynamics was a circle, in keeping with the group's preference for the free flow of information and ideas. The actual physical structure of the building (sometimes referred to as the Round House, or Calvin Carousel) was meant to promote interaction, discussion, and cooperation. Ideally, all working areas would be equidistant from the coffee pot situated on the large white central discussion table. However, the design was the source of great grief to the contractors, who had no right angles with which to fix the dimensions. Novel techniques were devised, and the construction was completed, with the official dedication ceremonies held on April 1, 1964. The gathering of funds for construction of the Laboratory of Chemical Biodynamics had begun as early as 1960. When I retired as director in 1980, the Regents renamed the building the Melvin Calvin Laboratory.

The circular layout of the laboratories and offices offered distinct advantages for interdisciplinary research. Scientists from several disciplines were brought together to share their different ideas and distinct approaches. The lab benches radiate outward like spokes from a wheel, and there is a spiral staircase, which rises behind the coffee table, to aid movement between the floors of the building. The design of the building was evolved so that the several scientific disciplines housed there could interact with each other and bring their various talents to bear on what we called "the dynamics of biological structure and function".

The Seminar. Our weekly seminars were begun about 1947 or thereabouts in the Donner Library and were held at eight

Gathered around the coffee table with UC president Clark Kerr (left) and Prince Philip (center), 1962. That table was the center of all laboratory activities, from actual coffee drinking to explaining the principles of two-dimensional paper chromatography to royalty.

o'clock on Friday mornings. The seminars were patterned after those in the chemistry department that were run by Gilbert N. Lewis many years earlier. In the beginning, and for many years thereafter, no assignments were made to speak at the seminar. Speaking was entirely impromptu, in the style that Lewis had originated. The way Lewis conducted the seminars was to have the faculty and students assemble in the seminar room, and then he would look around and ask someone to speak whom he had not heard for some time, or at least with whose work he was not familiar. I tried to emulate this and succeeded in doing so for many years. In other words, the group would assemble in the Donner Library, and I would look around and ask someone to tell us about his work. This approach had two merits: One was that it did not require previous preparation—there was no way to prepare for it except to be fully aware of what you were doing at all times; second, it was a help in the actual presentation because it was largely a matter of questions and answers rather than formal presentation. Any formal presentation was very quickly interrupted in order to find out what was really happening. This method gave rise to an exchange between the various members of the group in a way that nothing else could.

We maintained this style even after the laboratory was much larger and had a building of its own. Because of the great diversity of interests in the building, it finally became necessary to bring a little more "organization" to the weekly seminars. Schedules of speakers were drawn up, suggested by the senior staff members, and the person involved would receive a week's notice, or perhaps a little longer, to prepare his or her talk. This change in the method of operation was a result not only of diversity but sheer numbers, and it is the format of the seminars as they exist today.

Funding. In the early years in the Department of Chemistry, the research in which I was involved was sponsored by departmental funds obtained from the State of California and such private sources as the Petroleum Research Fund, Research Corporation, and the Rockefeller Foundation. The Rockefeller funds were especially important, as they could be used for much-needed equipment or to support foreign postdoctoral scholars; this support continued up to the 1960s. World War II changed forever the way science was supported in the university, when the government supplied funds through the Office of Scientific Research and Development (OSRD) and, of course, the Manhattan Project, establishing a precedent for large-scale support of research that continues today.

The Bio-Organic Chemistry Group that was started in 1945 has been funded by the government, first through the U.S. Atomic Energy Commission (initially arranged by E. O. Lawrence), then the Energy Research and Development Administration, and now the U.S. Department of Energy. In addition, other governmental agencies, such as the National Institutes of Health, and private industrial sources have supported the work of the group, largely on their own initiative.

Throughout the history of the laboratory, private sources of support have been invaluable. This was particularly true during the construction of the building, when one-third of the

In the Laboratory of Chemical Biodynamics in 1979. This view shows the arrangement of the laboratory benches and the coffee table, center, used for scientific discussion.

monies required was given by the Charles F. Kettering Foundation. This came about as the result of a request from the Kettering Foundation in early 1960 for me to come to their research laboratory in Antioch, Ohio, and review the work on behalf of their board. I went and spent a good deal of time listening to presentations of the research projects. I then wrote a report for them, as well as presenting my findings orally to the board.

Later that year, when the plans for our own building in Berkeley were developed, it became appropriate for me to seek funds, at least for the beginning of the building construction. My first thought was to telephone Carroll Hochwalt, President of the Monsanto Chemical Company and a member of the board of the Kettering Foundation, and tell him my problem. On the telephone his comment was, "Will $300,000 help?" I said, "Of course", and he said, "You have it." With that start, I then applied to the National Institutes of Health and the National Science Foundation for additional funds, and the final amount was supplied by the Regents of the University of California.

The reason for seeking outside funds for the building construction, even though the group had been supported since its inception by the U.S. Atomic Energy Commission (AEC), was that the AEC would not construct a building that it did not own and the Regents would not construct a building not their own to be located on the campus. If I had been willing to move

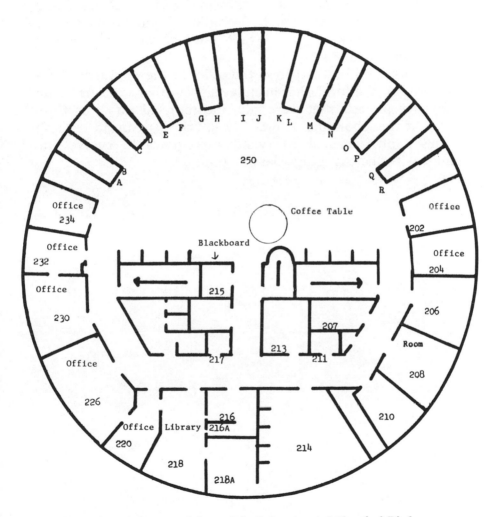

Floor plan of the second floor of the Laboratory of Chemical Biodynamics, showing the arrangement of work space.

the group up on the Hill with the other research groups that constituted what was then UCRL (University of California Radiation Laboratory), the AEC would have provided the construction funds. However, I was unwilling to do this because of my fear of being too separated from the interdisciplinary campus environment and the chemistry department.

Private support has also provided seed money over the years to begin new projects, such as the work on chemical and viral carcinogenesis and the hydrocarbons from plants project, allowing the work to go forward until government funding could be obtained.

During this entire period the office of the president of the University of California has made available other seed grants for innovative projects that were later incorporated into the mainstream of the laboratory effort. The quick response to requests of this nature enabled the various programs to evolve and change with the changing scientific priorities.

The Nobel Prize

Our work on the path of carbon resulted in the Nobel Prize in 1961. In retrospect, this event changed in some aspects the way my scientific life was conducted. Nobel Prize winners become "instant celebrities" and "instant authorities" for almost any type of endeavor, professional, political, or educational. We are besieged to lend our names to efforts of all types, and one of the most difficult things to do is to say "no" to many of these requests. Science goes on after the Prize, but the instant fame and interruptions can be overwhelming.

I first learned about the Prize at a classified meeting in Greenbelt, West Virginia, where there was a gathering of scientists of many different disciplines discussing the possibilities of using the large radiotelescope at Areceibo, Puerto Rico, to search for signs of extraterrestrial civilizations. Before I went to West Virginia there had been a week of wild speculation about the Prize, which had made life difficult for me, my family, and my personal staff. Access to communication at Greenbelt was extremely limited, but my wife did manage to get through to me with the news, and we had an immediate celebration. I left right away to return to San Francisco and to the celebrations that ensued during the next few days and weeks. The best way to describe what actually happened during this period leading up to the actual ceremonies in Stockholm is to quote from a letter that my wife wrote to one of her friends shortly after the Stockholm trip:

111

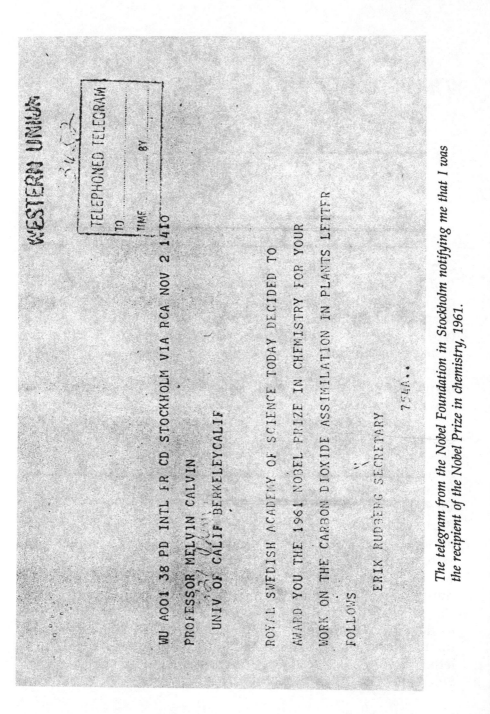

The telegram from the Nobel Foundation in Stockholm notifying me that I was the recipient of the Nobel Prize in chemistry, 1961.

The first information came to me by way of a phone call from NBC at 6:00 AM. Only after the official telegram was pried loose from the University mail by Melvin's secretary at about 10:00 AM was Melvin willing to concede that the announcement was authentic. . . . He arrived home at Midnight from his conference, and the next two days were filled with press conferences, a champagne party at the Faculty Club for lab people, a small party at home for old friends. And then he was off again for a week's lectures at Harvard!

There were, of course, more parties, shopping for formal clothes for the family, and innumerable incidents of humor and delight in the remaining weeks before departure. We (the entire family: Melvin, myself, daughter Elin (16); daughter Karole (13), and son Noel (8) flew by the SAS polar route to Copenhagen, with the intention of resting a few days before going on to Stockholm. But we hadn't realized that anonymity was no longer possible. We were anticipated or recognized everywhere, always with the greatest kindness and courtesy, but there were not nearly enough minutes for sleep.

Arrival in Stockholm was at 3:00 PM in the pitch dark, with a swirling snowstorm and fierce wind. There was a large delegation out in the snow to greet us.... The most detailed and well-considered arrangements had been made, tempered and refined by 60 years of making these occasions memorable. Dr. Erik Rudberg, the Director of the Nobel Foundation, was the official host, and the aide assigned to us from the Swedish Foreign Office...was exceptionally competent and, beyond that made every personal effort to make the visit comfortable for us all, particularly the children. In the entire visit, the wives and families of our hosts took as active a part as the hosts themselves.

From the moment of our arrival, a complete schedule had been made for every day, and the events of each day were all perfectly arranged. Of all the events, the three most outstanding were the For-

WESTERN UNION
TELEGRAM

W. P. MARSHALL, PRESIDENT

1201-(4-60)

The filing time shown in the date line on domestic telegrams is LOCAL TIME at point of origin. Time of receipt is LOCAL TIME at point of destination

0A360 =0

O WA363/GOVT PD=THE WHITE HOUSE 3 648P EST=

DR MELVIN CALVIN, DEPT OF CHEMISTRY=
UNIVERSITY OF CALIFORNIA BERKELEY CALIF=

:IT IS WITH GREAT PLEASURE THAT I SEND YOU MY SINCERE (
CONGRATULATIONS ON BEING THE RECIPIENT OF THE 1961 NOBEL
AWARD IN CHEMISTRY. THE PIONEERING RESEARCH YOU HAVE
CONDUCTED ON THE MECHANISM OF PHOTOSYNTHESIS HAS
REFLECTED MUCH CREDIT UPON YOU AND UPON THE DEVELOPMENT
OF SCIENCE IN THE WORLD TODAY. I WOULD LIKE TO ADD MY
BEST WISHES FOR THE CONTINUED SUCCESS OF THE IMPORTANT
INTERDISCIPLINARY RESEARCH WITH WHICH YOU ARE SO DEEPLY
CONCERNED=

JOHN F KENNEDY.

Congratulatory telegram from President John F. Kennedy.

mal Presentation of Awards, followed by the ball; the dinner given by the King and Queen at the Palace; and the Lucia festivities, including the Festival Ball given by the science students at the University.

The Nobel Ceremony is held in the Concert Hall. The Laureates, both past and present, are seated on the stage, and at the front of the audience, in specially arranged chairs, are seated the King and Queen and other members of the royal family. To their right and left are seated the members of the Laureates' families and the families of the officials of the Nobel Foundation, the Swedish Academy of Sciences, and the Royal Caroline Institute. The auditorium is banked with flowers, although this is deep winter, and television cameras and photographic equipment are so concealed by flowers that they are invisible. The music, both choral and instrumental, is beyond description, in both perfection of technical performance and in tonal depth.

The format of the ceremony remains the same, year after year, with the presentation of the Laureates and the awarding of the citations and medals by the King. It is interesting to note that for this ceremony the King sits below the Laureates, symbolizing the honor that he does to those chosen from the world, rather than from his own country. Speeches are beautifully done, in whichever language is suitable for the Laureate concerned, and translations into Swedish and English are provided.

The formal banquet is held in the gold-tiled assembly hall of the Stockholm City Hall. The appointments, food, and program are all beautifully planned, and the King keeps track personally of the minute-by-minute schedule. Following the banquet, dancing in the Blue Room, in which the students from the University participated, continues until about 2:00 AM.

Dinner at the Palace occurs on the following night. Although the social format is very thoroughly outlined, the formal structure is comparable to that of detailed rules in baseball.... So it is that the Swedish social customs appear formidable to an outsider, but once learned, there is a greater friendliness and ease of communication than at a western barbecue. The reason is clear, in that the formal rules have been developed so that no one is allowed to feel lonely, and each person present communicates with all the others. This is true at all levels of Scandinavian society.

As always throughout the festival, the flowers are exquisite, the food is peerless in quality, beauty, and service. Dinner in the Palace is less proscribed by the formal rules than in other places. Following the dinner, guests proceeded to the beautiful and ancient drawing room (a museum in itself) for conversation and refreshments. Promptly at 11:00 PM, after conversation with each special guest, the King and Queen take leave of their guests, and the party is over.

On December 13, Lucia Day in Sweden, we are serenaded in our hotel rooms at the Grand Hotel. The beautiful Lucia, usually the oldest daughter in the family, appears in white robes, with a crown made of greens and lighted by candles. She is attended by white-robed star boys, their headgear decorated with gold stars. The procession winds its way down the halls of the Grand Hotel and into all the Nobel rooms, singing in clear young voices "Santa Lucia, Santa Lucia". Lucia pours cups of coffee and serves star-shaped cookies for us. On the same day is the third party of special interest which is given by the science students at the University and is highlighted by the induction of the chemistry and physics Laureates into The Order of the Ever Smiling and Jumping Green Frog. It is a student party in its entirety, but it differs from its counterpart in the United States in that Swedish formality is ingrained in these young people.

...In retrospect, for the Laureates the most common factor (regarding the celebrations) is amaze-

Dr. Melvin Calvin, 1961 Nobel Prize winner in chemistry, with his wife, Genevieve, and family, Noel, 8, Karole, 13, and Elin, 15, board an SAS polar jet bound for Europe. Dr. Calvin's schedule includes a speaking engagement in Copenhagen at the Danish Chemistry Society and at four universities in the Scandinavian countries in addition to receiving his Nobel Prize in Stockholm.   —Scandinavian Airline photo

My family and I departing for Stockholm. (Photo courtesy of the Berkeley Daily Gazette, *December 8, 1961, and Scandinavian Airlines.)*

ment, in a most fundamental form, in an honor conferred on mankind through an individual. Most scientists recognize that they are at once a symbol of respect for all basic science, and that acceptance of this honor carries with it a responsibility for future conduct which will thereafter color all their lives.

Following the Stockholm festivities we visited (Melvin lectured) in Uppsala and Lund, Sweden;

Being presented with the Nobel Prize in chemistry from the King of Sweden, Stockholm, December 10, 1961, as my family looks on. The famous "Berkelely vest" (page 120) can be seen if one looks closely.

The press took note of the royal dinner partners at the Nobel gala. I am chatting with Princess Christina (top), and Genevieve with King Gustaf VI Adolph (bottom). (Photos courtesy of the Swedish–American News Service.)

Oslo, Norway; and Copenhagen, Denmark. All were delightful visits, each unique and each an interesting contrast to all the others. Seen in rapid and intimate succession, the shading of cultures from one area to another is quite evident and significant. One feels in these nations a tie with the past and with the future which gives every event a sense of history which both dignifies and deflates that individual's place in that panorama.

One other anecdote from the Nobel celebrations is of interest. Every university has its traditions, but it would be difficult to find one more impresssive than that launched by Edwin McMillan of Lawrence Berkeley Laboratory (LBL) in 1951. That was the year that McMillan and Glenn Seaborg traveled to Stockholm to accept the Nobel Prize in chemistry for their work on transuranium elements. McMillan purchased all the formal trappings for the ceremony, including a white vest. It was to become a familiar sight to Stockholmers. When Emilio Segre of LBL went over in 1959 to accept his Nobel Prize, he borrowed McMillan's white vest. Donald Glaser from Berkeley followed suit in 1960. By the time I made the trip in 1961, the Swedes were asking about the much-traveled Berkeley vest. Luis Alvarez of LBL also used it again in 1969 to continue the tradition.

We used the money from the Prize to purchase a ranch about 80 miles north of Berkeley in Healdsburg, Sonoma County. This has been our family retreat for more than 25 years. It consists of about 500 acres of rolling hills, with a creek meandering through and a large pond that we can use for swimming. For some years we kept horses there for riding, and we have continuously grazed cattle for other people on the land. The hillsides are too steep for planting grapes, even though the climate is ideal for them. The ranch is located in a mountainous area quite near an old mercury mine and also fairly close to the Geysers, one of the first thermal electrical generating plants in the United States. Because the ranch is so close to Berkeley, we have been able to go there frequently, and for many years the MC^2 Ranch, as it is called, was the site of the annual laboratory

HISTORIC OCCASION: The first dinner ever given in the White House to honor all living Nobel prize winners in the Western Hemisphere saw an exceptionally brilliant and distinguished gathering of guests last night. Above are some of those invited to one of the largest dinners ever given at the Executive Mansion. Front row, seated left to right: Mrs. Richard J. Walsh (Pearl Buck), Dr. Rudolf L. Mossbauer, Mrs. Ernest Hemingway, President Kennedy, Mrs. George C. Marshall, Dr. Melvin Calvin, Mrs. Kennedy and Dr. Robert Hofstadter. Standing directly behind the President, left to right: Dr. Isidor I. Rabi (wearing glasses), Dr. Thomas H. Weil F. Libby and Dr. Albert Szent-Gyorgyi.

By Vic Casamento, Staff Photographer

With my fellow laureates at the historic dinner given by the White House to honor all living Nobel winners in the western hemisphere. I am seated to the left of the First Lady. This event was the inspiration for President Kennedy's now famous and oft-repeated quote that "this is the most extraor-dinary collection of talent, of human knowledge...ever gathered at the White House, with the possible exception of when Thomas Jefferson dined alone." (Reprinted with permission from the Washington Post. Copyright 1962. Photographed by Vic Casamento.)

picnic. Also, as mentioned earlier, we used the ranch as a site for an experimental "plantation" of *Euphorbia lathyris* in order to develop agronomic techniques and to provide a seed source for our laboratory experiments.

One of the more spectacular events that we attended because of our status as Prize winners was a formal dinner at the White House in April 1962 when President and Mrs. John F. Kennedy had a special evening for all the Nobel Prize winners in the United States. In addition to the formal program of poetry read by Robert Frost, we danced with each other and met the Kennedys. I was particularly intrigued that evening to meet Pearl Buck, a person whose books I had long admired, and other laureates who were nonscientists. Apparently this event made quite an impression on the Kennedys as well, as their reaction to it was most enthusiastic.

The fact is, of course, that the recognition afforded by the Nobel Prize makes it somewhat easier for a scientist to obtain funding and, at the University of California in Berkeley, the ultimate prize: a personal, reserved free parking space (about 20 years later).

Service

My first introduction to federal government service occurred during World War II as a result of my association with the Manhattan Project and research support by the Office of Scientific Research and Development. After the war, when it was obvious that the support for basic science in the universities would come from the federal government, various requests would come for service to one group or another. One of the first of these was as a member of the advisory committee to the Chemistry Office of the Air Force (Air Force Office of Scientific Research), which was the basic research funding arm of that group.

One of the early international commitments, as a result of our experience with radioactive isotopes in the Donner and old Radiation Laboratory, was to join the Joint Committee on Applied Radioactivity (now the International Atomic Energy Agency), which consisted of representatives of various countries and which met twice a year to discuss problems of radioactivity and its distribution for research. At that time the various health hazards associated with radioactive materials were just being recognized, and one of our tasks was to ensure that adequate precautions were taken in the laboratory and in international shipments of materials. Other international committee service followed for the International Council of Scientific Unions, International Union of Pure and Applied Chemistry, and other groups of this type.

One of my longer standing commitments was to the National Aeronautics and Space Administration (NASA). I was a member of the Bioastronautics Committee, sponsored by the National Academy of Sciences Space Science Board, which actually was a forerunner of many NASA activities. This group, consisting of scientists of various backgrounds, actually met before NASA was established (as well as afterwards) to set up a priority system for work to be done on returned lunar samples as the Apollo Program developed. When the NASA effort was more formally organized, I was a member of the advisory committee for the returned lunar samples and planetary biology, an effort that lasted for an extended period. This group was responsible for the protocols for the returned lunar and planetary samples and how they were to be distributed and analyzed.

Simultaneously with the work on the advisory committee for lunar rock analysis, I became the U.S. chairman of a joint United States–Soviet effort to prepare a book on space biology and medicine. This effort, from the initial selection of authors to the final publication of the treatise, took 12 years, and it was the result of encouragement by the National Academy of Sciences, NASA, and the Soviet Academy of Sciences. Each chapter in the two-volume publication had an American and a Russian author, and this international cooperation involved yearly meetings of the various participants in either the Soviet Union or the United States. I personally believe that this was one of the more useful efforts in which I participated during my scientific life, as it spanned a period of time in which the relations between the Soviet Union and the United States became more open and our scientific contacts more fruitful. The people who prepared the various chapters were experts, and the book, which was printed in both English and Russian, has proved to be an important source of medical and biological information pertaining to space.

In 1962 I became a member of the Committee on Science and Public Policy (COSPUP) of the National Academy of Sciences. This was a group of scientists of different backgrounds, ranging from anthropology to zoology and even including economics (!), who met several times a year to develop policy and studies for scientific questions that had been given to the Academy for solution or policy suggestion. This association continued for many years, and I was chairman from 1972 to 1975. I found this service particularly stimulating. The members of COSPUP were outstanding scientists in their respective fields and members of the Academy. The interactions of the committee, the staff of the Academy involved with the committee, and various report committee members were very satisfying. The committee studied such questions as peer review, for example, in the early 1970s (a subject still under investigation) and helped prepare the first report on this subject. COSPUP met twice a year, once in Washington at the National Academy of Sciences and once in Berkeley at our laboratory. The Berkeley visits included social as well as scientific gatherings, and we always had the predinner cocktail parties at our home. This was a wonderful opportunity for the COSPUP members to meet the Berkeley laboratory staff who had worked so hard to make these gatherings successful, and to spend an evening of fellowship in one of the delightful eating establishments of the Bay Area.

Genevieve and I sharing a toast together in our home before the COSPUP party, 1977.

During the administration of President John F. Kennedy I was asked by George Kistiakowsky to become a member of the President's Science Advisory Committee (PSAC), a group established by President Eisenhower to advise him on matters that would affect science and that science itself might affect. There were representatives of all the physical sciences, engineering, and later also economics and political science. We met monthly and deliberated on various subjects, many of them military, preparing reports that were used by the government in planning. All of the members of PSAC had responsible positions in universities, federal laboratories, and industry and were able to conduct their efforts for PSAC with advice from the highest levels. After the death of President Kennedy in November 1963, I continued as a member of PSAC for some years during Lyndon Johnson's term of office. At the time I joined PSAC I knew most of the members personally, and these friendships have continued. Even today our contacts are frequent and rewarding.

Presiding over a meeting during my term as president of the ACS, 1970–1971.

Another long-term assignment was as a member of the Energy Research Advisory Board (ERAB). This group advised the Secretary of Energy and prepared in-depth reports on scientific proposals and broad research areas that were funded by DOE. These reports became benchmarks for development in many areas of research. The members of ERAB were chosen from various scientific backgrounds with experience in academic institutions, government laboratories, and industrial research organizations.

Throughout my scientific life in the United States I have been a member of the American Chemical Society; I served as Pacific Division Chairman in 1951. I was elected president of the national American Chemical Society for 1971, which required 3 years of intensive commitment as president-elect, president, and past president. At the time of my tenure, the chemical profession was in the midst of examining the relationship of the profession of chemistry to the employers (industry, government, and university) from the point of view of professional treatment. This was quite a controversial topic and involved discussions at all levels. This problem is still with the ACS today, and the resolutions and proposals have been defined at various levels for the chemical profession.

Industrial Consulting

Immediately at the end of the war in 1945 I was approached by David Pye, one of the scientists at the Dow Chemical Company Western Division in Pittsburg, California (formerly the Great Western Chemical Company), with the proposal that the Dow Company might be interested in the oxygen-generating systems that I had developed. This system depended on a family of cobalt chelate compounds. In the solid state these compounds could absorb oxygen out of the air with the liberation of heat, which was removed in a heat exchanger. After saturation, steam was introduced into the outside of the heat exchanger and pure oxygen was regenerated from the oxygenated solid cobalt chelates.

The Dow Company was interested in the synthesis of the salicylaldehyde derivatives that were involved in this process, as well as in the cobalt chelate itself. They invited me to visit their laboratory in Pittsburg, where I met the the director, Wilhelm Hirschkind, who had been the research director of the company when it was still Great Western Chemical Company. After a few visits, they invited me to become a general consultant in all areas of chemistry for the Pittsburg research laboratory in the Western Division, which I was pleased to do. After a few years of this kind of work, I was asked to visit the company headquarters in Midland, Michigan, and my consulting activities were thus expanded considerably. I then became a general consultant for the Dow Chemical Company in all its divisions and had occasion to visit not only the research laboratories in Midland but those in Freeport (Texas), Wayland (Massachusetts), and ultimately Dow-Lepetit (Milan, Italy), Dow-Zurich (Switzerland), Dow-Pacific (Hong Kong), and Asahi-Dow (Japan).

Later, after the company had established a manufacturing plant in Salvador, Brazil, we visited there as well. On one occasion, already mentioned, the Dow Chemical Company arranged an extensive tour for Genevieve and me throughout Brazil, beginning in Manaus on the Rio Negro, down the River Amazon to the bulge of Brazil to Salvador, and ultimately to the business office in Sao Paulo via Rio de Janeiro. This was a remarkable journey in which we had two young Brazilian employees of the Dow Company as guides. They had made the entire trip ahead of us, had checked out the hotels, boat trips, side trips of other types, etc., and they accompanied us on the trip itself. It was the kind of journey that happens only once in a lifetime and memories of it remain with us today.

Having become acquainted with all of the global research laboratories of the Dow Chemical Company and their various research directors, I was invited by Lee Doan, then chairman of the board, to become a member of the board of directors of the Dow Chemical Company. This was approximately in 1962 or 1963. There was a period of retirement from this activity in 1976 at the age of 65, which at that time was the mandatory age for retirement of the board members. However, after an absence of a year, the Dow board changed its bylaws to allow board members who had never been officers of the company to continue on until age 70, and I was invited back. The reason for this change was the mandatory legal requirement to have an audit committee of the board chaired by someone who was not an officer of the company and who had not been an officer of the company. For this reason I was asked to return and become chairman of the audit committee, a task for which I was not very well suited. In order to perform this task, because I had no previous experience with auditing procedures, I asked one of my colleagues here at the university who was dean of the School of Business Administration, Earl Cheit, to be my "consultant". He accompanied me to all of the audit committee meetings. We managed to survive them and to perform our duties acceptably until I finally did retire from the Dow board at the age of 70.

In addition to consulting for the Dow Chemical Company, in the middle 1950s I was invited to be a consultant for the Organic Chemical Division of the E. I. du Pont de Nemours research laboratory in Wilmington, Delaware; the Research Laboratory of the General Electric Company in Schenectady, New York; and the research laboratories of The Upjohn Company in Kalamazoo, Michigan. Each of these activities involved very different kinds of chemistry, and I found no overlap at all among them. Later on, however, a new research director, Roland Schmitt, was appointed by the General Electric Company to succeed Art Bueche. He was unaware of what I was actually doing for the General Electric research laboratory and felt there might be a conflict of interest because by that time I was a member of the board of Dow Chemical Company. Thus, my activities at General Electric were terminated. The Du Pont activities I terminated myself when I became a member of the Dow board, and the Upjohn connection was terminated when the Upjohn Company went into the heavy chemical business.

The Dow Chemical Company, Du Pont, and the Upjohn Company made substantial contributions annually to the university for my research activities. These contributions were allowed to accumulate over the years, were amalgamated into the General Endowment Pool, and still yield income to this day. The income from the fund, originally designated as the Director's Fund (for the Director of the Laboratory of Chemical Biodynamics) and now called the Calvin Fund, was (and still is) used for all sorts of special purposes in the laboratory for which federal monies were inappropriate.

There were two other large companies with which I had short-term connections. Both of these resulted from the growth of our interest in the production of hydrocarbons from plants, work done in collaboration with my wife, Genevieve. Thus, when we were invited to consult for the Diamond Shamrock Corporation and Estech, she was part of that consulting team.

More recently I have been connected with some new, smaller companies as a member of the scientific advisory boards that many new companies seem to feel are necessary. Most of them are concerned with chemicals from plants, and the connections with these companies as a member of their scientific advisory board still remain. Two others with which I am associated are Nova Pharmaceuticals (Baltimore, Maryland), which is involved in the development of drugs for the neurotransmitter and receptor systems, and Molecular Design, Inc., a Bay Area computer software designer particularly concerned with organic chemical problems.

Bringing It Together

The "umbrella" that brings together these seemingly unrelated scientific topics is the long-lived interest in coordination chemistry that began at the University of Manchester with Polanyi. The studies on phthalocyanine dyes led to interest in another coordination compound, chlorophyll, and eventually to the study of photosynthesis. The impulse to study photosynthesis also arose from another interest in coordination chemistry that appeared during World War II when we were involved in recovering uranium and plutonium from nuclear reactor materials. This method depended upon a coordination compound, which we designed and synthesized. The same scientific continuity extends to our work today in the area of artificial photosynthesis, where the redox chemistry of manganese complexes may lead eventually to the design of a device for the photolysis of water. One of the basic tenets of our scientific work has always been to go in whatever direction the "light" may lead—in this case, perhaps, the original flashlight was coordination chemistry and phthalocyanine.

However, there is even a broader-based guideline to what I have done. This was and is the willingness to undertake work to learn whatever facts and techniques might be necessary to answer the scientific questions that came my way. Although this attitude may have some intrinsic foundation, there is no question in my mind that the examples of my two most influential

In my office, October 1991.

teachers, Michael Polanyi and Gilbert Lewis, played a major role in nurturing it and bringing it to fruition. Thus, I found myself working at some time or another in atomic physics, physical chemistry, organic synthesis, cellular biochemistry, neurochemistry, plant molecular biology, chemical evolution and organic geochemistry, biophysics, and animal behavior. These were not conscious choices to change direction; they were simply the consequences of necessity.[99]

In one of my lectures many years ago I used the phrase "following the trail of light". The word "light" was not meant in its literal sense, but in the sense of following an intellectual concept or idea to where it might lead. My interest in living things is probably a fundamental motivation for the scientific work in the laboratory, and we created here in Berkeley one of the first and foremost interdisciplinary laboratories in the world. Perhaps my philosophy may best be expressed in the following statement: "There is no such thing as pure science. By this I mean that physics impinges on astronomy on the one hand, and chemistry and biology on the other. The synthesis of a really new concept requires some sort of union in one mind of the pertinent aspects of several disciplines. . . . It's no trick to get the right answer when you have all the data. The real creative trick is to get the right answer when you have only half of the data in hand and half of it is wrong and you don't know which half is wrong. When you get the right answer under these circumstances, you are doing something creative". This is really what the Calvin laboratory was all about.

The Beautiful and Awesome Order

Just before the final draft of this book was typed in 1987, Genevieve Calvin died, another victim of cancer. We had been married for almost 45 years, and my personal life was built around her presence. Whatever I did, I did in the light of her possible reaction to it in my consciousness.

She was a genuine collaborator with me in my scientific endeavors, ranging from the work on the Rh incompatibility factor through the joint efforts on the use of hydrocarbon-producing plants as alternatives to fossil fuels for the production of energy and materials. Especially important were her contributions to the philosophy of science, and the articles that I prepared in this vein were the result of her insight and collaboration.[100] Her name appears with mine as a joint author on nine scientific papers, and she independently contributed to educational publications, especially those oriented toward youth and the place of the family in society, as well as environmental journals.

Genevieve's presence was felt throughout the scientific community in which we live and work. People all over the world were her friends and colleagues, and her wisdom and insight into scientific and human problems that she shared with them was of the highest order. After 1950 she accompanied me on all of my trips abroad, including sabbatical leaves, sometimes with members of our family. Once we visited her parents' family home in Norway, and we also spent a sabbatical at Oxford

University (1967–1968) when I was George Eastman professor there. She attended the international meetings where I spoke, and very often participated in the scientific sessions as well as the many social events that accompanied gatherings of this type. In fact, on several occasions she presented scientific papers of her own authorship at international meetings. These papers were on the subject of plants as alternatives to fossil fuels for energy and materials production. She truly was a world citizen. Gen also took part in the many activities required of me as a member of the board of directors of the Dow Chemical Company and enriched her worldwide circle of friends through these contacts.

She made our house a "home" to the visiting foreign postdoctoral people who came to the laboratory in Berkeley, and many times they spent their first few nights in Berkeley in our cottage. She made the annual event of the picnic and reception for the laboratory members unique and special, and did heroic service particularly at the time of the Nobel Prize. The memories many of our former colleagues have of wonderful times in Berkeley were enhanced by her concern for all of us.

I would like to end with one of her poems, which was written for a commencement address at the College of Wooster, Ohio,[101] in 1966.

Respect thyself, lest thy respect be
         unworthy to give.

Order thy days, lest thy loved ones
         be ensnared
         in thine own confusion.

Honor the wisdom of thy heart by the quiet
         of the hour of its borning.

Reflect, for thou art a part
         of all men,
         of all life,
         of all matter
         of all of the stars and
         the voids of eternity.

Then only can the
         beautiful and awesome order
         which is your being
         give to your span of days
         that small new goodness, or
         wisdom, or
         beauty
         which determines tomorrow.

                    Genevieve Calvin

Epilogue

Celebrations and Reminiscences

The Calvin Laboratory Reunion

In February 1989 there was a reunion of members of the Calvin Laboratory from the 1940s, 1950s, 1960s, and 1970s. The genesis of the get-together came about through a phone call received by my secretary, Marilyn Taylor, in Boulder, Colorado, at the home of Professor Bert Tolbert, one of the original senior staff of the Bio-Organic (now Calvin) group. His brother, Professor Nathan Tolbert of Michigan State, also a Calvin Laboratory alumnus, happened to call, and these three thought a scientific reunion was a splendid idea. When Marilyn returned to Berkeley, the planning began, involving three other senior people in Berkeley from the Calvin Laboratory: Richard M. Lemmon (retired), James Alan (Al) Bassham (retired), and Marie Alberti. This scientific and social reunion was the result of 2 years of intensive effort. (One reminder is found in the two lineal feet of files that contain the history of the event, including the financial records!) Perhaps if they had known what was involved in such an event, they would not have been so eager to start!

More than 60 scientists came to Berkeley from all over the United States and foreign countries (England, Switzerland, Belgium, and New Zealand, to name but a few). In addition to the usual social events, a scientific symposium was held, with alumni

February 3, 1989

Dear Calvin-Reunion Participant:

```
Welcome!      Welcome!      Welcome!
```

Our schedule for the day is as follows:

12 noon-1:30 pm Meet for lunch in Room 116 of the Calvin Laboratory. Enjoy conversations with old/new friends. Be overcome with nostalgia as you examine the historical memorabilia.

1:30 pm Welcome to LCB and brief summary of our current research programs by acting Director, John Hearst.

1:40 pm (a). **TOURS of LABORATORY of CHEMICAL BIODYNAMICS** led by John Hearst and Jim Bartholomew.
(b). **VISITS** with friends, see other memorabilia (Room 226).
(c). Start of **VIDEOTAPING** sessions in Room 116. Anyone interested is welcome to watch these informal group talks. Participants, and approx. times are as follows:

1:40 pm PHOTOSYNTHESIS
Al Bassham (discussion leader), Melvin Calvin, Clint Fuller, Murray Goodman, John Hearst, Mel Klein, Rod Park, Ken Sauer, Nate Tolbert

2:20 pm Tour of LCB for photosynthesis group participants. This tour will be videotaped.

3:00 pm ISOTOPIC SYNTHESES
Bert Tolbert (discussion leader), Melvin Calvin, Al Bassham, Dick Lemmon, Winifred Tarpey, Pete Yankwich

3:30 pm ANIMAL BIOCHEMISTRY
Ed Bennett (discussion leader), Ann Hughes, Ann Orme, Hiromi Morimoto, Marie Alberti, Bill Byrne

4:00 pm CANCER RESEARCH
Jim Bartholomew (discussion leader), Mina Bissell, Ercole Cavalieri, Joe Landolph

4:30 pm Adjourn

The schedule of activities for the Calvin reunion.

The 1940s group at the 1989 reunion. Front row, left to right: Marilyn Taylor, James Bassham, Bert Tolbert, me, and Murray Good-man. Back row: Richard Lemmon, Peter Yankwich, Nathan Tolbert, Lorel Kay, and Edward Bennett.

discussing the four broad areas of science that have always been studied in the Calvin Laboratory: synthetic organic chemistry and reaction mechanisms (including chemical evolution), animal biochemistry, carcinogenesis (chemical and viral), and photosynthesis. The symposium was also videotaped for the LBL (Lawrence Berkeley Laboratory, formerly the Radiation Laboratory, Rad Lab) archives; individuals who wished a copy of any of the tapes were able to purchase them.

A celebratory dinner was held at the Faculty Club, the scene of Calvin Laboratory celebrations over the years. It was an outstanding evening, and one of the guests, Andrew Benson, arrived from the Seychelles, having flown nonstop to get there. Speeches, toasts, photographs, anecdotes, and laughter were captured on audio tape, which was transcribed and a copy sent to all the participants. Throughout the 2-day event, roaming photographers and video cameras were evident to capture the proceedings of the symposium and the camaraderie of the reunited group. Class pictures of those in the Calvin Laboratory during the various decades were taken and sent to those lucky enough to come. The organizing committee could not believe the alacrity of positive response to a chance to come to Berkeley to celebrate the outstanding scientific and human experiences that they have had in the Calvin Laboratory. In many instances colleagues were reunited who had not visited Berkeley in more than 30 years.

Accolades

Another important event occurred in 1989. I was one of the recipients of the 1989 National Medals of Science. The ceremony at the White House and the other social events were also attended by my older daughter, Elin.

In 1991 I traveled to Washington again to receive the John Ericsson Medal from the Department of Energy. This was in recognition of our work on a program of renewable energy sources such as hydrocarbon-producing plants and artificial photosynthesis. This time I traveled with my younger daughter, Karole, and my 12-year-old grandson, Kris. This trip was a memorable scientific and family activity for all of us.

To Melvin Calvin
with best wishes, Ge Bush

Receiving the prestigious National Medal of Honor from President
George Bush at the White House, 1989.

Two types of parties celebrating the Nobel prize were
given at our house in Berkeley in 1961. The first was a rather
small dinner for approximately 30 people who were most closely
associated with the work at Berkeley. Included were my
mother, Rose Calvin, and my mother-in-law, Karen Jemtegaard,
as well as Mrs. Molly Lawrence (wife of the first LBL director),
Edwin and Elsie McMillan (he was director of LBL at the time),
and Helen and Glenn Seaborg, the dean of the College of Chemistry. The laboratory members who attended this special party
had worked closely together for many, many years, and all felt
personally that they were sharing the glory and excitement. The
other Nobel party was a much larger and less personal event,

Back in Berkeley the Nobel Prize celebrations continued. Here are my mother and I at a dinner given at my home, November 1961. My mother made the tie I am wearing; it has a "photosynthesis" motif. I have a collection of unique ties (one displays the DNA chains), some gifts from colleagues and many handmade.

again at the house, but the guest list was increased to more than 100, including all the graduate students, postdoctoral associates, and laboratory employees and their spouses and friends. Everyone who was there that night will always remember the atmosphere, the food, the fun, the camaraderie.

A Birthday to Remember

An 80th birthday is one of life's mileposts and, therefore, it seemed appropriate to recognize the day—April 8, 1991. As with the Nobel prize, I was feted with two types of celebrations. The first, a small gathering in my office, actually occurred on my birthday. The birthday group were those who are involved in

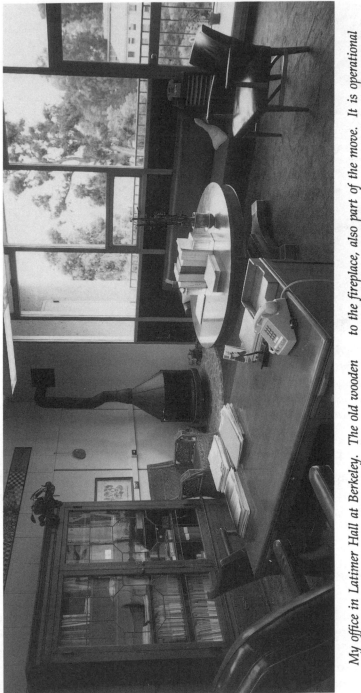

My office in Latimer Hall at Berkeley. The old wooden bookcases with their leaded-glass doors were from my office at the old Chemistry building, which was razed in 1963. The Escher print above the bookcase leads the eye to the fireplace, also part of the move. It is operational and is used frequently in the winter. I eventually bought the furniture, which was once in my laboratory.

the current research on artificial photosynthesis, with the addition of Professor Glenn T. Seaborg, a close personal friend since I came to Berkeley in 1937. At that time, as bachelors, we lived at the Faculty Club, and we have maintained close personal ties throughout the intervening years. Our careers followed parallel paths in many ways, as both of us came to Berkeley while Gilbert Lewis was alive,[98] both of us became associated with Ernest Orlando Lawrence and his fledgling Radiation Laboratory, and both of us received the Nobel Prize in Chemistry, Glenn in 1951, and I, 10 years later in 1961. One more recent parallelism occurred when Professor Seaborg won the National Medal of Science in 1991.

At the office gathering, with suitable libation of champagne accompanied by strawberries, Seaborg and I reminisced about the early days at Berkeley. Both of us were associated with Gilbert Lewis, the outstanding Dean of the College of Chemistry. Seaborg received his Ph.D. in 1937 at Berkeley, and Lewis didn't recommend him for a position anywhere. Actually, that was a good sign, and, as Glenn said,

> It meant I still had a chance to stay on at Berkeley, which, of course, was my ambition. One day Lewis called me [Seaborg] into his office . . . and he asked me if I would like to be his research assistant. I said, "Who, me?" I was completely flabbergasted and asked if he was sure I could actually fill this role. He indicated that if he didn't think so, he would not have given me the opportunity. . . . When I finally completed my 2-year term as Lewis's personal assistant, I remember that he called me into his office. I was also doing a great deal of other work at the same time in the Radiation Laboratory. Lewis informed me that he was terminating my appointment as his research assistant and putting me on the faculty as an instructor.[98]

At the party the postdoctoral associates and others who had not lived through those magnificent years enjoyed the opportunity to listen to us two members of the "Rad Lab Gang" reminisce about earlier (and simpler?) days at Berkeley. We are quite different persona: Seaborg is well over 6 feet tall, very thin, always dressed in a three-piece suit; I'm about 5 feet 7 inches, bald, with brown eyes, and usually wearing a tweed jacket and turtleneck shirt. We're probably one of the campus "sights" when we walk together to the Faculty Club for lunch. Glenn had recently (February 1991) been honored at a special symposium celebrating the 50th anniversary of the first isolation of plutonium. We talked about the war days in Berkeley and told anecdotes about colleagues then and now. Glenn reminded me that my laboratory was also on the same floor of Gilman Hall as the room where the actual isolation of plutonium took place. However, many in the Chemistry Department really were not too aware of what was going on on the third floor of Gilman because of the secrecy of the project.

The partiers enjoyed listening to us chat about what seemed to many a totally unknown era, peopled by sometimes irrepressible scientists, rushing around the country on trains, involved in research that was crucial to the birth of the atomic age. We were two friends, whose days together had gone back more than 50 years, so full of interest and enthusiasm not only for what was then but for what is now, and whose birthdays are both in April.

The second and truly important celebration of my 80th birthday took place at my home on Sunday, April 14, 1991. This house had been the scene of more exciting gatherings than could be counted since 1951 when we bought it. The house was designed by the eminent Berkeley architect, Bernard Maybeck, and is one of those special places that is almost magical in feeling and absolutely individual in design. For example, included on the grounds is an eight-sided music studio, as the original owner(s) were involved in musical performance and wanted a private place for practicing! Our home was the locale for parties large and small—the yearly dinner for the postdoctoral associates and graduate students, the Calvin Nobel festivities in Berkeley in 1961, dinners for various National Academy committees, faculty dinners and parties, and festivities for the Dow Board of Directors.

Most who came to the 80th birthday celebratory dinner this year had been to the house many, many times, and it

*Daughter Elin Sowle at my
80th birthday party.*

*Daughter Karole Campbell was also on hand for my 80th birthday cele-
bration.*

Granddaughter Meadow Sowle.

Grandchildren Kris and Robin Campbell.

seemed auspicious to come there again. There had been very little partying at 2683 Buena Vista Way for the last several years, since Genevieve's death, and the house almost seemed to open its arms to friends and colleagues. The group that gathered were a close-knit family. The attitude of close cooperation in the laboratory carried over to close cooperation socially. These people had not only partied together but had hiked the Sierras in the summer, skied together in the winter, and jostled for symphony and opera tickets in the fall.

The party was planned by my two daughters Elin and Karole. Three of my grandchildren, Meadow (age 18), Robin (age 17), and Kris (age 11) were present and gave a feeling of continuity to the gathering. Many of the guests had known my daughters when they were children; association with the Calvin family by the senior people was especially close. The food and wine were wonderful, and a beautiful string quartet played quietly from an upstairs balcony. Whimsical gifts were opened to great peals of laughter and applause. After the birthday cake and candle blowing-out, we again began to talk about early days in Berkeley, what the Chemistry Department and LBL were like in the "old days". One story led to another, and some of us who thought we had heard everything found out that there was a great deal more to know.

It turned out that there were four present at the birthday dinner who had been involved with the Calvin Laboratory continuously since the early 1940s. One (Bert Tolbert) is now a retired professor of chemistry from the University of Colorado at Boulder; two (Dick Lemmon and Al Bassham) are retired staff senior scientists from LBL who were graduate students in 1945–1946; and one was my secretary, Marilyn Taylor. The four told "how we got to the Calvin group in the first place" tales, ranging from being in the right place at the right time (the retired professor from Boulder who had received his Ph.D. at Berkeley in 1947 and was asked to take on administrative chores of the Calvin group in 1945); the two graduate students, one of whom (Dick Lemmon) had been told by his research professor at the California Institute of Technology to come to Berkeley and work with Melvin Calvin, and the other graduate student (Al Bassham) who happened to learn of our photosynthesis work

With Marilyn Taylor.

and became one of the early students in that area; and the secretary.

In those ancient days, being hired by the Rad Lab as a secretary involved a long process, including vetting for a security clearance; there were many weeks between job interview and date of employment, weeks fraught with insecurity and impatience. Marilyn Taylor, the secretarial candidate, was interviewed by four very bright Ph.D. group leaders before meeting me. (She remembers when she was an undergraduate, seeing me running through the Old Chemistry Building with a very dirty laboratory coat on the way to class.) After many conversations with the four Ph.D.s and lots of discussion about "How long do you plan to work here", "Do you plan to get married, and when", "If you get married do you plan to have children?" (the group had lost two secretaries to pregnancy), we finally met. We talked a couple of minutes, and we both said "You'll do". Marilyn has "done" ever since.

References

1. Glockler, G.; Calvin, M. *J. Chem Phys.* **1935**, *3*, 771; Glockler, G.; Calvin, M. *J. Chem. Phys.* **1936**, *4*, 492.

2. Calvin, M. *Trans. Faraday Soc.* **1936**, *32*, 1428; Calvin, M.; Dyas, R. E. *Trans. Faraday Soc.* **1937**, *33*, 1492.

3. Calvin, M., Cockbain, E. G.; Polanyi, M. *Trans. Faraday Soc.* **1936**, *32*, 1426.

4. Calvin, M.; Eley, D. D.; Polanyi, M. *Trans. Faraday Soc.* **1936**, *32*, 1443.

5. Calvin, M. *J. Chem. Educ.* **1984**, *61*, 14.

6. Calvin, M. *Trans. Faraday Soc.* **1938**, *34*, 1181.

7. Calvin, M. *J. Am. Chem. Soc.* **1939**, *61*, 2230.

8. Calvin, M.; Wilmarth, W. K. *J. Am. Chem. Soc.* **1956**, *678*, 1301.

9. Halpern, J. *J. Organomet. Chem.* **1980**, *200*, 133.

10. Calvin, M. Welch Foundation Conference on Chemical Research, Robert A. Welch Foundation: Houston, **1976**, *20*, 116.

11. Lewis, G. N.; Calvin, M. *Chem. Rev.* **1939**, *25*, 273.

12. Lewis, G. N.; Calvin, M. *J. Am. Chem. Soc.* **1945**, *67*, 1232.

13. Lewis, G. N.; Calvin, M.; Kasha, M. *J. Chem. Phys.* **1949,** *17,* 804.

14. Scheibe, G. *Angew. Chem.* **1939,** *52,* 631.

15. Jelley, E. E. *Nature* **1935,** *139,* 631.

16. Branch, G. E. K.; Calvin, M.; *The Theory of Organic Chemistry;* Prentice-Hall: New York, 1941.

17. Hammett, L. P.; *Physical Organic Chemistry: Reaction Rates, Equilibria and Mechanisms;* Mc-Graw Hill: New York, 1940.

18. Calvin, M.; Kodani, M.; Goldschmidt, R. *Proc. Natl. Acad. Sci. U. S. A.* **1940,** *26,* 340.

19. Calvin, M.; Kodani, M. *Proc. Natl. Acad. Sci. U. S. A.* **1941,** 27, 291.

20. Pauling, Linus; *The Nature of the Chemical Bond and the Structure of Molecules and Crystals: An Introduction to Modern Structural Chemistry;* Cornell University Press: Ithaca, NY, 1940.

21. Pfeiffer, P. *Ann.* **1933,** *503,* 84.

22. Tsumaki, T. *Bull. Chem. Soc. Jpn.* **1938,** *13,* 583.

23. Calvin, M.; Bailes, R. H.; Wilmarth, W. K. *J. Am. Chem. Soc.* **1946,** *68,* 2254; Barkelew, C. H.; Calvin, M. *J. Am. Chem. Soc.* **1946,** *68,* 2257; Wilmarth, W. K.; Aronoff, S.; Calvin, M. *J. Am. Chem. Soc.* **1946,** *68,* 2263; Calvin, M.; Barkelew, C. H. *J. Am. Chem. Soc.* **1946,** *68,* 2267; Hughes, E. W.; Wilmarth, W. K.; Calvin, M. *J. Am. Chem. Soc.* **1946,** *68,* 2273; Harle, O. L.; Calvin, M. *J. Am. Chem. Soc.* **1946,** *68,* 2612; Bailes, R. H.; Calvin, M. *J. Am. Chem. Soc.* **1947,** *69,* 1886.

24. Reid, J. C.; Calvin, M. *J. Am. Chem. Soc.* **1950,** 72, 2948.

25. Martell, A. E.; Calvin, M.; *The Chemistry of the Metal Chelate Compounds;* Prentice-Hall: New York, 1952.

26. Moskowitz, M.; Dandliker, W. B.; Calvin, M.; Evans, R. S. *J. Immunol.* **1950,** *65,* 683; Evans, R. S. Moskowitz, M.; Calvin, M. *Proc. Soc. Exp. Biol. Med.* **1950,** *73,* 637; Dandliker, W. B.; Moskowitz, M.; Zimm, B. H.; Calvin, M. *J. Am. Chem. Soc.* **1950,** *72,* 5587.

27. Calvin, M.; Evans, R. S.; Behrendt, V.; Calvin, G. *Proc. Soc. Exp. Biol. Med.* **1946,** *61,* 416.

28. Ruben, S.; Hassid, W. Z.; Kamen, M. D. *J. Am. Chem. Soc.* **1939,** *61,* 661.

29. Weigl, J. W.; Calvin, M. *J. Chem. Phys.* **1949,** 17, 210.

30. Calvin, M.; Heidelberger, C.; Reid, J. C.; Tolbert, B. M.; Yankwich, P. E. *Isotopic Carbon;* Wiley: New York, 1949.

31. Tolbert, B. M.; Adams, P. T.; Bennett, E. L.; Hughes, A. M.; Kirk, M. R.; Lemmon, R. M.; Noller, R. M.; Ostwald, R.; Calvin, M. *J. Am. Chem. Soc.* **1953,** *75,* 1867; Lindblom, R. O.; Lemmon, R. M.; Calvin, M. *J. Am. Chem. Soc.* **1961,** *83,* 2484.

32. Lemmon, R. M.; Mazzetti, F.; Reynolds, F. L.; Calvin, M. *J. Am. Chem. Soc.* **1956,** *78,* 6414.

33. Bassham, J. A.; Calvin, M. *The Path of Carbon in Photosynthesis;* Prentice-Hall: Englewood Cliffs, NJ, 1957.

34. Bassham, J. A.; Calvin, M. *The Photosynthesis of Carbon Compounds;* W. A. Benjamin: New York, 1962.

35. Calvin, M. *Science* **1962,** *135,* 879.

36. Consden, R.; Gordon, A. H.; Martin, A. J. P. *Biochem. J.* **1944,** *38,* 224.

37. Calvin, M. *J. Chem. Educ.* **1958,** *35,* 428.

38. Calvin, M.; Dorough, G. D. *J. Am. Chem. Soc.* **1948,** *70,* 699.

39. Huennekens, F. M.; Calvin, M. *J. Am. Chem. Soc.* **1949,** *71,* 4024, 4031; Seely, G. R.; Calvin, M. *J. Chem. Phys.* **1955,** *23,* 1068.

40. Calvin, M.; Buckles, R. E. *J. Am. Chem. Soc.* **1940,** *62,* 3324.

41. Splitter, J. S.; Calvin, M. *J. Org. Chem.* **1958,** *23,* 651; *J. Org. Chem.* **1965,** *30,* 3417; *Tetrahedron Lett.* **1968,** *1445; Tetrahedron Lett.* **1970,** 3395; Splitter, J. S.; Su, Tah-Mun; Ono, H.; Calvin, M. *J. Am. Chem. Soc.* **1971,** *93,* 4072; Ono, H.; Splitter, J. S.; Calvin, M. *Tetrahedron Lett.* **1973,** *42,* 4107.

42. Calvin, M.; Barltrop, J. A. *J. Am. Chem. Soc.* **1952,** *74,* 6153; Barltrop, J. A.; Hayes, P. M.; Calvin, M. *J. Am. Chem. Soc.*

1954, *76,* 4348; Calvin, M. *Fed. Proc.* **1954,** *13,* 697; Whitney, R. B.; Calvin, M. *J. Chem. Phys.* **1955,** *23,* 1750.

43. Kearns, D. R.; Calvin, M. *J. Chem. Phys.* **1958,** *29,* 950; Tollin, G., Kearns, D. R.; Calvin, M. *J. Chem. Phys.* **1960,** *32,* 1013; Kearns, D. R.; Tollin, G.; Calvin, M. *J. Chem. Phys.* **1960,** *32,* 1020.

44. Kearns, D. R.; Calvin, M. *J. Am. Chem. Soc.* **1961,** *83,* 2110; *J. Chem. Phys.* **1961,** *34,* 1022.

45. Garrison, W. M.; Morrison, D. C.; Hamilton, J. G.; Benson, A. A.; Calvin, M. *Science* **1951,** *114,* 416.

46. Simpson, G. G. *The Meaning of Evolution;* Yale University Press: New Haven, CT, 1949.

47. Miller, S.L.; Urey, H. C. *Science* **1953,** *117,* 528; Miller, S. L. *J. Am. Chem. Soc.* **1955,** *77,* 2351; Miller, S. L.; Urey, H. C. *Science* **1959,** *130,* 245.

48. Calvin, M. *Am. Sci.* **1956,** *44,* 248; *Chem. Eng. News* **1961,** *39(5) (May 22),* 97–104. *Proc. Roy. Soc.* **1965,** *288A,* 441.

49. Ponnamperuma, C. A.; Lemmon, R. M.; Bennett, E. L.; Calvin, M. *Science* **1961,** *134,* 113; Ponnamperuma, C. A.; Lemmon, R. M.; Calvin, M. *Radiat. Res.* **1963,** *18,* 540; Ponnamperuma, C. A.; Lemmon, R. M., Mariner, Ruth; Calvin, M. *Proc. Natl. Acad. Sci. U. S. A.* **1963,** *49,* 735.

50. Ponnamperuma, C. A.; Lemmon, R. M.; Calvin, M. *Science* **1962,** *137,* 605.

51. Steinman, G.; Lemmon, R. M.; Calvin, M. *Proc. Natl. Acad. Sci. U. S. A.* **1964,** *52,* 27; Schimpl, A.; Lemmon, R. M.; Calvin, M. *Science* **1965,** *147,* 149.

52. Steinman, G.; Kenyon, D. H.; Calvin, M. *Nature* **1965,** *206,* 707; *Biochim. Biophys. Acta* **1966,** *124,* 339.

53. Calvin, M. *Science* **1959,** *130,* 1170; *Perspect. Biol. Med.* **1962,** *5,* 147.

54. Calvin, M. *Angew. Chem. Int. Ed. Engl.* **1974,** *13,* 121.

55. Warden, Jr., J. T.; McCullough, J. J.; Lemmon, R. M.; Calvin, M. *J. Molec. Evol.* **1974,** *12,* 481.

56. Allen. D. E.; Gillard, R. D. *J. Chem. Soc. Chem. Commun.* **1967,** *1091.*

57. Havinga. E. *Biochim. Biophys. Acta,* **1954,** *13,* 171.

58. Calvin, M. *AIBS Bull.* **1962,** *12(5),* 29. *Adv. Biol. Med. Phys.* **1962,** *8,* 315; *Chem. Br.* **1969,** *1,* 22; *Am. Sci.* **1975,** *63,* 169.

59. Eglinton, G.; Calvin, M. *Sci. Am.* **1967,** *216(1),* 132; Calvin, M. *Perspect. Biol. Med.* **1969,** *13,* 45.

60. Calvin, M.; Vaughn, S. K. In *Space Research;* Kallman, H., Ed.; North–Holland: Amsterdam, 1960; pp 1171–1191.

61. Eglinton, G.; Scott, P. M.; Belsky, T.; Burlingame, A. L.; Calvin, M. *Science* **1964,** *145,* 262.

62. Burlingame, A. L.; Haug, P.; Belsky, T.; Calvin, M. *Proc. Natl. Acad. Sci. U. S. A.* **1965,** *54,* 1406; Van Hoeven, W.; Maxwell, J. R.; Calvin, M. *Geochim. Cosmochim. Acta* **1969,** *33,* 877; Han, J.; Calvin, M. *Nature* **1969,** *224,* 57.

63. Han, J.; McCarthy, E. D.; Van Hoeven, W.; Calvin, M.; Bradley, W. H. *Proc. Natl. Acad. Sci. U. S. A.* **1968,** *59,* 29; Han, J.; McCarthy, E. D.; Calvin, M.; Benn, W. H. *J. Chem. Soc.* **1968,** *C,* 2785; Han, J.; Chan, H. W.-S.; Calvin, M. *J. Am. Chem. Soc.* **1969,** *91,* 5156; Han, J.; Calvin, M. *Proc. Natl. Acad. Sci. U. S. A.* **1969,** *64,* 436; *J. Chem. Soc. Chem. Commun.* **1970,** 1470; Philp, R. P.; Brown, S.; Calvin, M. *Geochim. Cosmochim. Acta* **1978,** *42,* 63.

64. Philp, R. P.; Calvin, M.; Brown, S.; Yang, E. *Chem. Geol.* **1978,** *22,* 207; Philp, R. P.; Calvin, M. In *Advances in Organic Geochemistry,*1975; Ensidema Publishing Co.: Madrid, Spain. 1977; pp 735–752.

65. McCarthy, E. D.; Calvin, M. *Nature* **1967,** *216,* 242.

66. Burlingame, A.L.; Calvin, M.; Han, J.; Henderson, W.; Reed, W.; Simoneit, B. R. *Science* **1970,** *167,* 751; Burlingame, A. L.; Calvin, M.; Han, J.; Henderson, W.; Reed, W.; Simoneit, B. R. *Geochim. Cosmochim. Acta* **1970,** *2(1),* 1779; Henderson, W.; Kray, W. C.; Newman, W. A.; Reed, W. E.; Simoneit, B. R.; Calvin, M. *Geochim. Cosmochim. Acta* **1971,** *2(2),* 1901.

67. Calvin, M. *Chemical Evolution: Molecular Evolution Towards the Origins of Living Systems on Earth and Elsewhere;* Oxford University Press: New York, 1969.

68. Calvin, M.; Sogo, P. B. *Science* **1957**, *125*, 499; Sogo, P. B.; Pon, N. G.; Calvin, M. *Proc. Natl. Acad. Sci. U. S. A.* **1957**, *43*, 387.

69. Calvin, M. *Rev. Mod. Phys.* **1959**, *31*, 147, 257; Androes, G. M.; Calvin, M. *Biophys. J.* **1962**, *2*, 217.

70. Eastman, J. W.; Androes, G. M.; Calvin, M. *J. Chem. Phys.* **1962**, *36*, 1197; *Nature* **1962**, *193*, 1067.

71. Calvin, M. *J. Theor. Biol.* **1961**, *1*, 258.

72. Hughes, A. M.; Tolbert, B. M.; Lonberg-Holm, K.; Calvin, M. *Biochim. Biophys. Acta* **1958**, *28*, 51.

73. Hughes, A. M.; Calvin, M. *Science* **1958**, *127*, 1445; Hughes, A. M.; Bennett, E. L.; Calvin, M. *Proc. Natl. Acad. Sci. U. S. A.* **1959**, *45*, 581.

74. Lewis, G. N. *Science* **1934**, *79*, 151.

75. Calvin, M.; Hermans, Jr., J.; Scheraga, H. A. *J. Am. Chem. Soc.* **1959**, *81*, 5048.

76. Heidelberger, C.; Brewer, P.; Dauben, W. G. *J. Am. Chem. Soc.* **1947**, *69*, 1389.

77. Cavalieri, E.; Calvin, M. *Photochem. Photobiol.* **1971**, *14*, 641; Joss, U. R.; Calvin, M. *J. Org. Chem.* **1972**, *37*, 2015; Tischler, A. N.; Thompson, F. N.; Libertini, L. J.; Calvin, M. *J. Med. Chem.* **1974**, *16*, 227; *J. Med. Chem.* **1974**, *17*, 948.

78. Meehan, T.; Warshawsky, D.; Calvin, M. *Proc. Natl. Acad. Sci. U. S. A.* **1976**, *93*, 1117; Meehan, T.; Straub, K.; Calvin, M. *Proc. Natl. Acad. Sci. U. S. A.* **1976**, *93*, 1437; Meehan, T.; Straub, K.; Calvin, M. *Nature* **1977**, *269*, 725; Straub, K.; Meehan. T.; Burlingame, A. L.; Calvin, M. *Proc. Natl. Acad. Sci. U. S. A.* **1977**, *74*, 5285.

79. Calvin, M.; Joss, U. R.; Hackett, A. J.; Owens, R. J.; *Proc. Natl. Acad. Sci. U. S. A.* **1971**, *68*, 1441; Hackett, A. J.; Owens, R. B.; Calvin, M.; Joss, U. R. *Medicine* **1972**, *51*, 175; Calvin, M. *Radiat. Res.* **1971**, *50*, 105; Hackett, A. J.; Sylvester, S. S.; Joss, U. R.; Calvin, M. *Proc. Natl. Acad. Sci. U. S. A.* **1972**, *69*, 3653.

80. Calvin, M. In *Viral Replication and Cancer*; Melnick, J. L.; Ochoa, S.; Oro, J., Eds.; Editorial Labor S.A.: Barcelona, Spain, 1973, pp 195–218; Calvin, M. *Naturwiss.* 1975, 62, 405; *Prog. Biochem. Pharmacol.* 1978, 14, 6; In *The Role of Chemicals and Radiation in the Etiology of Cancer*; Huberman, E., Ed.; Raven Press: New York, 1985; pp 51–64.

81. Hughes, A. M.; Calvin, M. *Cancer Lett.* 1976, 2, 5; 1979, 6, 15; Hughes, A. M.; Tenforde, T. S.; Calvin, M.; Bissell, M. J.; Tischler, A. N.; Bennett, E. L.; *Oncology* 1978, 35, 76.

82. (a) Calvin, M. *Science* 1974, 184, 375; (b) *Am. Sci.* 1976, 64, 270; (c) *Photochem. Photobiol.* 1976, 23, 425; (d) *Energy Res.* 1977, 299. (e) *BioScience* 1979, 29, 533; (f) *Discover* 1981, 1 . (g) *Science* 1983, 219, 24 (h) *J. Appl. Biochem.* 1984, 6, 3; (i) *Ann. Proc. Phytochem. Soc. Eur.* 1985, 26,147; (j) *Bot. J. Linn. Soc.* 1987 , 94, 97.

83. Adams, R. T.; McChesney, J. D. *Econ. Bot.* 1983, 37, 207.

84. Nielsen, P. E.; Nishimura, Hiroyuki; Otvos, J. W.; Calvin, M. *Science* 1977, 198, 942.

85. Nemethy, E. K.; Otvos, J. W.; Calvin, M. *J. Am. Oil Chem. Soc.* 1979, 56, 957; *Pure Appl. Chem.* 1981, 53, 1101; In *Fuels from Biomass*; Klass, D. L.; Emert, G. H., Eds.; Ann Arbor Science Publishers: Ann Arbor, MI, 1981, pp 405–419.

86. (a) Calvin, M. *Naturwiss.* 1980, 67, 525; (b) Calvin, M.; Nemethy, E. K.; Redenbaugh, K.; Otvos, J. W. *Experientia* 1982, 38, 18.

87. Nemethy, E. K. *CRC Crit. Rev. Plant Sci.* 1984, 2, 117.

88. Taylor, S. E.; Calvin, M. *Comments Agric. Food Chem.* 1987, 1, 1.

89. Calvin, M. *Acc. Chem. Res.* 1979, 11, 369; *Energy Res.* 1979, 3, 73; Photochem. Photobiol. 1983, 37, 349; *J. Membrane Sci.* 1987, 33, 137.

90. Ford, W. E.; Otvos, J. W.; Calvin, M. *Nature* 1978, 274, 507; *Proc. Natl. Acad. Sci. U. S. A.* 1979, 76, 3590.

91. Tributsch, H.; Calvin, M. *Photochem. Photobiol.* 1971, 14, 95.

92. Okuno, Y.; Ford, W. E.; Calvin, M. *Synthesis* 1980, 7, 537; Willner, I.; Otvos, J. W.; Calvin, M. *J. Chem. Soc. Chem. Commun.* 1980, 964.

93. Laane, C.; Ford, W. E.; Otvos, J. W.; Calvin, M. *Proc. Natl. Acad. Sci. U. S. A.* **1981**, *78*, 2017; Laane, C.; Willner, I.; Otvos, J. W.; Calvin, M. *Proc. Natl. Acad. Sci. U. S. A.* **1981**, *78*, 5928.

94. Willner, I.; Otvos, J. W.; Calvin, M. *J. Am. Chem. Soc.* **1981**, *103*, 3203; Willner, I.; Yang, J.-M.; Laane, C.; Otvos, J. W.; Calvin, M. *J. Phys. Chem.* **1981**, *85*, 195; ACS Symposium Series 177; American Chemical Society: Washington, DC, 1982; p 71.

95. Loach, P. A.; Calvin, M. *Biochem.* **1963**, *2*, 361; *Biochim. Biophys. Acta* **1964**, *79*, 374; Calvin, M. *Rev. Pure Appl. Chem.* **1965**, *15*, 1.

96. Spreer, L. O.; Maliyackel, A. C.; Holbrook, S.; Otvos, J. W.; Calvin, M. *J. Am. Chem. Soc.* **1986**, *108*, 1949; Spreer, L. O.; Leone, A.; Maliyackel, A. C.; Otvos, J. W.; Calvin, M. *Inorg. Chem.* **1988**, 27, 2401; Brewer, K. J.; Liegeois, A.; Otvos, J. W.; Calvin, M.; Spreer, L. O. *J. Chem. Soc. Chem. Commun.* **1988**, 1219; Brewer, K. J.; Calvin, M.; Lumpkin, R. S.; Otvos, J. W.; Spreer, L. O. *Inorg. Chem.* **1989**, *28*, 4446.

97. Grant, J. L.; Goswami, K.; Spreer, L. O.; Otvos, J. W.; Calvin, M. *J. Chem. Soc. Dalton Trans.* **1987**, 2105; Craig, C. W.; Spreer, L. O.; Otvos, J. W.; Calvin, M. *J. Phys. Chem.* **1990**, *94*, 7957.

98. Calvin, M.; Seaborg, G. T. *J. Chem. Educ.* **1984**, *61*, 11.

99. Calvin, M. *Photosyn. Res.* **1989**, *21*, 3.

100. Calvin, M. *CERN Courier* **1962**, 2, 12; Calvin, M.; Calvin, G. J. *Proc. Am. Phil. Soc.* **1964**, *51*, 73; Calvin, M.; Calvin, G. J. *Am. Sci.* **1964**, *52*, 163.

101. Calvin, G. J.; Calvin, M. *Main Currents in Modern Thought;* **1967**, *23* (3), 59.

Index

Index

Copy editing: Colleen P. Stamm
Production: Peggy D. Smith
Indexing: A. Maureen Rouhi
Production Manager: Robin Giroux

Printed and bound by Maple Press, York, PA